After a successful career in journalism, Mike Tomkies returned to his childhood love of nature and spent over thirty years living in remote places in the Scottish Highlands, Canada and Spain. During this time, he wrote a number of best-selling books, including *Alone in the Wilderness*, *A Last Wild Place*, *Out of the Wild*, *On Wing and Wild Water*, *Wildcat Haven* and *Moobli*. He died in 2016.

A Last
Wild
Place

Seasons in the Wilderness

MIKE TOMKIES

Introduction by Jim Crumley

This edition published in Great Britain in 2021
under license from Whittles Publishing by
Birlinn Ltd
West Newington House
10 Newington Road
Edinburgh
EH9 1QS

www.birlinn.co.uk

ISBN: 978 1 78027 703 5

British Library Cataloguing-in-Publication Data

A catalogue record for this book is available on
request from the British Library
Papers used by Birlinn are from well-managed
forests and other responsible sources

Typeset by Hewer Text UK, Edinburgh
Printed and bound by Clays Ltd, Elcograf S.p.A.

Contents

PART FIVE

PART SIX

Introduction

With the publication of the first edition of *A Last Wild Place* in 1984, Mike Tomkies had finally found his feet in the West Highlands of Scotland, and found his nature writer's voice. His earlier Scottish books, *Between Earth and Paradise* and *Golden Eagle Years*, seem in retrospect to have had the air of a work in progress, rather than the finished article.

At the heart of *A Last Wild Place* are fifteen chapters centred on the loch, woodlands, and mountains that surrounded the isolated cottage he called Wildernesse, each landscape element portrayed and examined through all the seasons of the year.

'Over the years,' he wrote, 'I have learned it is upon these stages, through the pageantry of all four seasons, the interplay of day and night, that nature plays all the mysteries, dramas, comedies and tragedies of creation itself. Often I go, admission free, a humbled audience of one, to watch and wonder.'

In *A Last Wild Place*, Mike Tomkies took the reader into his confidence in a way that no-one else had done before. Seton Gordon, for example, unarguably the most significant of his predecessors among what he called 'the high and lonely places', was a naturalist and writer of great gifts, but there was a quality of aloof restraint about his writing. In contrast, Mike Tomkies wanted his readers to engage with him, and they did, in their thousands. His way was to show them not just what this looks like and sounds like, but what it *feels* like to be here on this mountain at this moment. He also ensured that they knew what it takes to produce such work, the sheer physical slog that underpinned a quite extraordinary dedication to fieldwork in such a landscape.

I had first met him in 1982 when I was a newspaper journalist on the *Edinburgh Evening News* and interviewed him when he was promoting *Golden Eagle Years*. I subsequently visited him at Wildernesse, and over the ensuing years we became good friends. But it was when I helped him on treks to eagle eyries that I gained a healthy respect for what he was trying to do there.

When *A Last Wild Place* appeared, I reviewed it for the *Edinburgh Evening News*. If you were to take my original copy down from my bookshelves none too carefully, a piece of paper would fall out. It's a neat and flawlessly typed half-page letter from Mike, thanking me for the review and my understanding of his work, and wishing me well in my declared intentions to give up my staff job and become a nature writer.

Re-reading the book now, more than thirty years after I did just that, I remember the impact it made on me, and its author's generosity and encouragement in the face of my own early struggles. But I am just one of many, many people who owe a debt of gratitude to Mike's work, to his work ethic, and to this book in particular. The example it sets for its uncompromising pursuit of nature's secrets is a lesson thoughtfully taught and well-learned. It remained his own favourite of all his books throughout his writing life, and I am delighted that it lives on in this new edition.

The final chapter of *A Last Wild Place* is titled 'Renewal'. Mike writes:

In nature's teeming world the animals and birds are working hard to fulfil their destinies. The feeling came strongly upon me that we, who evolved from original creation to become the dominant species . . . have an inherent and inescapable duty to act as responsible custodians of the whole inspiring natural world. We are the late-comers, it can only be ours on trust.

. . . If we fail to learn from the last wild places, we may yet create a hell on earth before we too pass along the road to extinction, the fate of all dominant species before us . . . The lessons will not wait for ever to be learned.

It was true in 1984. It is surely even truer today as our planet warms and our climate lurches into chaos.

When Mike Tomkies died in October 2016, nature mourned the loss of a friend and an ambassador to its cause.

Jim Crumley
Stirling, 2017

PART ONE

Chapter 1

Into the Wild

The old stone cottage was cold and cheerless. As I shivered on a fish box in its empty dampness for the first time, surrounded by the debris of my former life, I felt a sudden panic. Soaked to the skin, I was fatigued after a long day ferrying boat loads of lumber and belongings up the steely grey loch under the onslaught of heavy rain. Now, my cherished hope that after seven years of wilderness living, first on a remote coast in western Canada and then in an old wooden croft on the Atlantic edge of a Scottish island, this new place would be my base for the deepest wilderness experiences of all, seemed not only presumptuous but foolhardy. No one had lived here, all year round, since 1912. When the old steamer that had once plied the loch was removed, so had died the human life along its shores.

Wildernesse, the name I had already given to the old cottage and the two small woods that flanked it, stood below a deep cleft in the Inverness-shire mountains in one of the largest uninhabited areas left in the British Isles. It was the only surviving dwelling in fifteen miles of roadless loch shore, and while I was used to isolation it would clearly be the wildest, loneliest place in which I had ever lived.

As I had found out on camping treks before leasing it, behind my new home lay a sixty-square-mile trackless wilderness of misty valleys and rivers, wooded ravines and gorges; and the first 500-foot ridges that rose steeply behind the cottage were merely the start of a succession of jagged granite outcrops and rolling hills that plunged into deep glens and led to a range of mountains whose peaks, snow-capped much of the year, towered to almost 3,000 feet.

Up there, somewhere, roamed herds of wild red deer whose stags, roaring in the autumn rut, would soon be dominating the landscape; a kingdom where golden eagles soared the skies, sharing their mountain domain with buzzards, ravens, kestrels and ptarmigan; where wildcats had their dens in rocky cairns, and at night badgers plodded from their setts, and the hill foxes roamed for miles. Across the loch to the south lay an area just as wild but three times the size.

Now, although my new domain had seemed a potential wild paradise at first view, the actual day of the move had shown me it might turn out the very opposite.

My doubts had begun at dawn when I had woken to find that two weeks of sunny mid-September skies had given way overnight to wintry darkness and drizzle, so the last boat trip from the sea island that had been my home for nearly four years had made the wrench of leaving a doubly miserable affair. No sooner had I manhandled the 500 lb weight of my new 15-foot 6-inch boat down over some pine branches into the fresh water loch that was to be my new home water and loaded it to the gunwales, than the weather worsened. The drizzle turned to sheets of cascading rain, the wind increased and as I steered fearfully through the rolling waves the long miles to Wildernesse it seemed as if the new loch was determined to make me pay a high price for moving into its world.

For the rest of the day I and my island friend Iain MacLellan, who had been given a one-day-only permission to drive my heaviest gear and old boat on a coal lorry to a private track on the loch's south shore for the move, laboured to ferry everything across in sousing rain. We carried up more than fifty loads, struggling together with an old calor gas fridge, a mahogany desk, lumber and a heavy twenty-foot wooden scaffold device I had invented for roof repairs. At last, as dusk approached, we faced each other, totally drenched, our forelocks dangling like rats' tails, and Iain bade me a cheerful goodbye.

'Well, 'tis yourself for the hardy life now. Sooner you than me!'

Alone again, as I heaved more gear out of the rain, I was acutely aware that my nearest neighbour lived over six miles away to the west. From now on my only link with the outside world, through the winter storms ahead, would be my small boat. Just to get mail (the post office had told me not even telegrams could be delivered) or supplies I could not grow for myself, I faced a thirteen-mile return boat trip, to where I parked the Land Rover in a small pine wood. Or else a long hike with hefty backpack over rock-filled bogs, high ledges, fallen trees in the long steep woodlands, and through snagging bracken, slushy peat and areas of ankle-wrenching tussocks whose flowing grassy heads hid the gullies between them.

To camp out alone in the wilds for a few days or to spend a summer holiday in such isolation was one thing, but now I faced the rest of my life, all the year round, in this lonely spot. No electricity, phone, TV, piped gas, not even a road.

A kind of madness, it would seem to many, but I had long since come to terms with my penchant for the last wild places. Initially, after years of journalism in many of the leading cities of Europe and America, I had fled to the strangeness of British Columbia to write the Great Novel. The book had failed but in those Pacific coastal wilds a city hedonism had been exorcised, the love of nature that I had discovered as a boy in Sussex had been reborn, and finally I had trekked into the last remote fastnesses to watch grizzly bears, cougars, bald eagles and caribou in the wild. Then I had become bewitched by a desire to try and also live a wilderness life back in Britain. Perhaps obeying some deep ancestral calling – for my mother had been a Highlander – I had indeed found it on the sea island of Shona, off the Inverness-shire coast. There, learning to live partly from the land and sea, had come my first close experiences with creatures like red deer, foxes, wildcats, sparrowhawks, herons, sea birds and seals. Yet those years served only to whet my appetite and now my love for the wildlife of the Highlands, still one of

Europe's finest wild regions, had all the cravings of an uncon-summated love affair.

Although my earlier romantic notions about nature had long been knocked out of me, I had become consumed by a passion to sink myself totally into one of Britain's last wild places, to live simply, close to the vital animal state myself the full year round in an even wilder place of higher mountains, longer rivers and larger forests, amid greater remoteness, this time in a fresh water environment. Only then, I felt, by more persistent study, could I possibly understand the vast interplay of Highland nature, from strand of moss to massive oak, from tiny beetle and little wren to mighty stag and golden eagle, and perhaps communicate through my writings not only love for wildlife but the necessity of conserving, even enhancing, the inspiring natural world, for the sake of man himself. To communicate my own conviction that unless man redeems the heedless greed that has destroyed so much of that world he may yet destroy the very environment he needs for both physical and spiritual suste-nance, and that he, once the brightest light of evolution, will end up its greatest failure.

Pretentious aims they were, of course, but there was adventure in it all too, and I confess that after a travelling varied life, noth-ing else now seemed more worthwhile than to pass on what most captivated my own heart and spirit.

Well, I thought as I sank a needed dram and wiped my drip-ping hair with a towel, now I have the chance – if I survive!

I looked around. Twenty-four sacks of belongings lay in defeated attitudes against each other on the stone floors. Old friends of books with sodden covers seemed to stare up at me with reproach for subjecting them to even greater hardship in their old age. The mahogany desk given to me by Shona's owners was in its future place in the fourteen-foot-square western room that was to be my study. From its window I could still see in the gathering dark a gleaming panorama of four miles of turbulent loch before it vanished in a bend between the hills – merely the last leg of the day's hard drenching journeys. In the kitchen of similar dimensions that lay at the other end of the cottage beyond a small bedroom, the heavy old calor gas fridge, the first preserver of provisions I had owned in seven wilderness years, stood against a cold stone wall. Outside the wooden scaffold rested on the roof, ready for its repairs. At least the heaviest items were in place.

Wearily I went out into the rain again and down the muddy forty-yard path to the shore and brought up two more loads. Then my body gave out. Too tired to peel vegetables, I warmed up a tin of soup, spooned it down from the can and, oblivious to the rain still thundering on the tin roof, collapsed into my sleeping bag.

During the days following, as I toiled to spruce up the cottage and build furnishings to suit the future life style, I was to live in a state of anxious sensitivity to my surroundings. Every unusual incident assumed the importance of an omen of good or ill for the future, each moment of wild beauty glimpsed eclipsed by something that caused fearful doubt.

When I went down next morning, rain still teeming in heightened south-west gales, I was horrified to find both boats, which I had hauled up on to wooden planks, were swamped and banging about in the waves. But for the intervening floating lumber, much of it now scattered a hundred yards along the lochside, they would have been smashed up against the rocks and destroyed. Now I was living by fresh water for the first time I had imagined that at last I would not have to cope with the tides of the sea, but the incessant rain had caused the loch to rise two feet overnight. As I laboured knee deep in water to bail them out faster than the waves filled them, the loch seemed to be trying maliciously to wreck my boats, as if angry at this interloper who did not follow a traditional pursuit like crofting with sheep. Then I realised these waters and the cradling hills had existed long before the coming of the sheep which caused the dispossession of so many Highland clans in the 1790s.

While each wave crashed in as if personally directed, I flailed away with a bucket until I had both boats floating again so I could haul them out. Some of the items I had left on the bank – cement, plaster, candles, tins of paint – had been scattered along the shore. I had to laugh then, for each time I had moved into a new wild place I had lost my first bag of cement in some such way. I should have remembered and carried it up before running out of strength. This time I decided to foil the nemesis and use it before it set hard.

Feverishly I mixed in fine gravel, then dashed about the cottage with bucketfuls, filling holes in the walls and climbing up the wooden scaffold and about the rain-slippery roof to patch up the chimneys. Prodding the cottage's foundations with a crowbar, I found about two cubic feet of stone and old mortar on the south-west corner had decayed. I chopped the rot away, sledgehammered in more squarish rocks, trowelled the last cement into the crevices then smoothed it all off. I resisted the temptation to crow over cheating the loch. I had long learned that wild places can take a nasty revenge on newcomers who get too uppity – and at Wildernesse I could sometimes feel a presence, as if I were being watched by the spirits of those who had gone before.

It was no use sorting out the impossible piles of gear until the cottage's inner walls were decorated so for the rest of the morning I shovelled flaking plaster from the study walls and put on fresh. I painted them white, to give more light from the single window. That afternoon the rain ceased, the light improved and while I only had the place on a twenty-year lease I walked round the new land with a feeling of proud possession. I now saw its full beauty as the trees still held their summer foliage.

West of the cottage stood a four-acre conifer wood filled with scaly red-barked Scots and other pine trees, silver and Douglas firs, with small ash trees, rowans, oaks, birches and holly in the clearings. A thirty-foot escarpment rose from its centre, leading to areas of mushy swamp, tussocks and hugh boulders cushioned with moss. East of the cottage and right next to it was a triangular wood of largely deciduous trees, great spreading oaks and tall spires of larches, dominated by a huge Norway spruce that was a full fifteen feet round the butt. Linking these giants were dozens of nut-bearing hazel bushes, more ashes, rowans, hollies, birches and occasional windbreaks of evergreen rhododendrons. A grove of stately beeches flanked the burn which ran down the edge of the wood and was the main vein of the mountains for over seven miles. A hundred yards northeast of the cottage the burn flowed over four deeply stepped pools, dropped down a ten-foot fall into the pool that held my water pipe, then cascaded in three separate forks thirty feet on to a tangle of massive gnarled rocks. Little ferns sprang from the fairy grottos in this spectacular waterfall and in the sunshine myriad rainbows were formed from the spray. The two woods framed Wildernesse and with the protection of the hills that rose immediately behind gave it a unique mini climate of its own.

Between the cottage and the loch, itself framed by a fringe of alders and ashes along the shore, was a 1½-acre patch, now almost totally engulfed in six-foot-high bracken. Well, I would clear this wild tract by hand scythe, not by chemicals, though it would mean cutting the virulent weed back several times a year, to increase the wild flowers and food plants for many insects, bees, and beautiful butterflies and moths. From it too I would wrest a vegetable patch. But right now I had to sort out the interior of my little dwelling.

When I found an uprooted larch tree leaning over at an alarming angle and creaking in the wind I had an idea. Among the lumber I had towed over was a huge plank of red pine, fourteen feet long, a foot wide and four inches thick, riddled with decorative teredo worm boreholes, that had drifted up on a sea beach at Shona. From this I could make a mantelpiece above the fireplace in the study. I also needed two stout logs to frame the fireplace and hold up the mantel, and the backs of these logs would need to be flattened, laboriously with an axe, to fit against the wall.

As I watched the fifty feet of the fourteen-inch-diameter larch swaying up and down, I thought that with a little care I might use the weight of the tree to crack the backs off the logs I needed. I had only an old logging handsaw and with this I perspiringly cut two-thirds through the larch about twelve feet above the roots. Sure enough, there was a sharp report – and a two-foot crack appeared neatly under the cut. I then made a similar cut four feet nearer the roots, to give me the first log. The crack widened until it almost reached the saw. I made a third cut another four feet nearer the roots but by now the top of the tree had come to rest on its branches on the ground. These I axed away one by one and the tree sank further, the split catching up with the saw. With a few sideways axe strokes against the last fibres I had my two four-foot logs, each with a third of its width split off right down its back. I de-barked them with a hand axe, peeled them until they gleamed yellow, and soon had my fireplace's log frame and massive mantel in place.

By nightfall I had knocked together from fish boxes and other lumber two sets of bookcases and camera shelves to flank each side of the fireplace. I found by cutting the side planks of the shelves at a slight angle they tilted into the wall and the weight of the books held the whole lot glued there, by gravity.

During the following days I had no time for debilitating doubts about the rightness of my move but took refuge in dawn to dusk work. I painted the chimneys with red oxide, replaced some rusty sheets of corrugated iron on the roof, renewed washered nails, wirebrushed off rust, and slapped costly green bitumen paint – to blend better with the landscape – over the entire fourty-foot roof until my left arm shook with fatigue as it propped up my weight on the scaffold and my right flailed away with the brushes. I cut the rest of the pine slab into two four-foot lengths and made a solid workbench for future carpentry. From some piping framework found on a rubbish dump in London, and more odd lumber, I made a record and battery player 'coplex', then put up more bookshelves, kitchen shelves, and fashioned a working area near the two-burner camp cooker from an old discarded table top. I decorated the other walls, painted the wooden ceilings after brushing them with woodworm fluid, and cut and hauled firewood for the coming winter. As I only had the handsaw for the first years, I used log hauling and cutting like a city man uses visits to a gym – as an antidote to hours at the writing desk. Thus it became less of a chore, and I soon had the old woodshed behind the cottage half full of fuel.

One problem was the big boat. I didn't want to haul its 500 lb weight over planks or logs every time I returned from supply trips, which would also wear out its hull. So with fourteen-foot planks and 2-inch by 2-inch poles as 'rails' held apart at the right distance to fit the rubber wheels of the trolley by 4-inch by 2-inch struts, I made two portable (just about) runways. Although the loch level rose a few feet after rainstorms one runway could be put into the water at the right depth, the boat could be slid on to the trolley, then winched up to the end and on to the second runway. Then the first runway could be put at the end of the second, the boat winched up further, and so on until I had it a safe distance from the loch.

To leave the boat in the water at the pine wood where I parked the Land Rover on supply trips I evolved a slip rope system. I obtained a heavy anchor with a galvanised metal ring on its top. To this I tied a pulley. With two ropes through the ring and one through the pulley I thus had three twelve-yard-long ropes from the bow that joined up with nearly a hundred yards of single rope tied to the stern. When I reached the far end I dropped anchor, paddled stern first to shore, hopped off with my gear, then with gentle tugs hauled the boat bow-first back out to above the anchor before tying the rope to a pine root. This way I had three ropes holding the boat to the anchor and even in storms, provided I didn't leave it too long, it was unlikely all three would chafe through.

The advantage of both systems was that I never needed an unsightly mooring buoy, nor a permanent structure for hauling the boat out, to blight the wild landscape. When the time came to depart again I could, like a nomadic Indian or Arab, do so without leaving any trace.

It was my first supply trip that brought back doubts as to the wisdom of my move. I hit into two hailstorms, the second squall so thick with the falling, bouncing stones I had to throttle down because I could see no further than ten feet. But it soon passed and I reached the pine wood safely. The trouble was that just to get mail and supplies, for I also had to drive a further seven miles to the village, took five and a half hours.

It was a rough trip back too and when I arrived at my beach in the near dark, I could not clearly see the trolley and runways. Waves crashing on shore made it impossible to use them anyway. I turned the bow into the surge at the last second, leaped off, and, using each bang of a wave as a natural helping shove, hauled her out on to planks. So fast had these movements to be I slipped on the rocks in the dark and gashed one hand.

The following day dawned in the first bright sky since my move and helped restore some optimism for the future. I had just finished painting one outside wall when my gaze fell upon the calm shimmering loch. To hell with work, I thought, I'm going fishing. It was rather late in the year and all I knew about Scottish fresh water loch fishing then was that you could troll an artificial spinning lure slowly behind the boat for salmon. I had done this kind of fishing in Canada and still had my lures – so out I went in the small boat.

The sun was high in the sky and the molten glassy surface of the loch mirrored the tiny stunted firs and pines on some nearby islets. Great dark cormorants, clinging to rocks with reptilian feet, dried their jagged wings in the sun. I hoped that meant there were plenty of fish in the loch. As I headed east, the mountains on either side rose precipitously from the shores, perfectly match-ing their reflections which blurred and vanished as the small bow wave rode over them. The slopes of the upper glens and corries far ahead were clothed in many shades of green, like velvet cushions thrown down by the hand of God, and as little clouds moved overhead, single shafts of sunlight burst through to illumine them one by one, as if each were saying: 'Now, now. Look at my beauty next.' In one small inlet white water lilies swayed in a slight breeze above their flat leathery leaves, anchored by snaky stems, and when I looked down the water was so clear, the bottom so far below, I felt as if I were magically suspended from a height.

Over all hung a haze of a golden blue harmony in which boat and human seemed to move in an enchanted dream, and the sense of space and wildness was overpowering and sweet. I caught no fish but just to be there was food enough for the spirit.

It was the next day, the last of September, that my doubts returned. I woke to find a white mist lying like a wreath over the loch, separating me from the rest of the world. 'But,' I wrote defiantly in my diary, 'I care less and less for the world and what people may think of me. There were aeons when man didn't exist and there'll be aeons more when he's gone. And when nature returns to take over the blemished land, nothing much will have happened. And yet – I am in the same trap. Just hope I can get my work done before I go.'

As the mist cleared, sucked up suddenly as if by a giant vacuum, it revealed the hills behind in stark blazing light. Up there lay that wildlife kingdom whose secrets I had come all this way to know. It was time to start the *real* work, I decided. I put camera, lenses, film, notebook, and a piece to eat into my backpack, my fieldglass into a top pocket of my camouflage jacket, and set off on the steep climb northwards.

At first I headed up the burn itself, surprised after a half mile at the depths of the gorges through which the waters had carved their way over thousands of years. Where the sides were sheer I climbed the slippery rocks above the flowing water, then headed among the stunted trees of the less steep sides, warily skirting sheer drops of thirty or forty feet. There were places here where no sun ever shone yet the foliage was lush, with spleenworts and thick lichens with flabby leaves that hid the white bark of the birches. The air was damp and noisy and filled with spray mist below the many falls.

On small ledges tiny ashes, alders and willows grew where no sheep or deer could get at them. But not a single bird or animal did I see. I criss-crossed the chasms until I emerged beyond the first high ridges. Here the land opened into broad meadows, spiked with sudden rock faces and filled with deep tussocks where I learned to keep close to the slow tawny waters of the burn that now meandered more lazily through the shallower country. Here were narrow belts of bright green grassland along the banks where the soil, being well drained and following occasional limestone strata, was more fertile than the wet peaty acid ground that spread out on either side. The oval droppings of deer showed how much they favoured these small oases for grazing.

After two miles I came to an isolated black tarn set in a broad shallow bowl of black peat at about 1,600 feet but not a fish disturbed its surface, and when I looked into the clear water there seemed no sign of insect life. I headed on across the expanse of undulating hills and as far as I could see there was no sign of man either. It was wild, bare and open, like a moon landscape, and I felt few men could ever have set foot there. I walked on between the tussocks until I came to a ridge overlooking a great glen stretched out far below, and then, on the side of it way down low I saw the bracken covered undulating furrows of some ancient 'lazy beds' where potatoes had once been planted, and below them a few lost stones of a tiny dwelling. It was hard to believe small crofting communities had once used these wild places, taking their cattle high and sleeping out with them during their summer grazings. Across the far side of the glen reared the twin cones of mountains whose peaks nearly 3,000 feet high poked above a ring wreath of clouds.

Almost four miles I had climbed and walked and I had seen no wildlife whatever, certainly not an eagle, nor even a buzzard, raven or crow. What kind of country is this? I wondered. How can I study the wildlife of these hills if there isn't any? How much wilder, wider, bigger and more empty they seemed than the punier hills of Shona island. How hard would I have to work to study *anything* here? I had a fearful feeling that at forty-five I had come to this wild place too late, that the ability to cover twenty miles a day of such hills on foot was over for me. What had I done? What foolish move had I made?

I turned, feeling downhearted, but just as I did so I saw some brown lumps on a hill slope over a mile away. I looked through the fieldglass. At last! There was a huge herd of red deer hinds, calves and yearlings. Some were sitting down as if chewing the cud while others were grazing and I counted forty-seven but as some walked on to the skyline I realised this was not the whole herd and there must be even more on the other side.

I was cheered by the sight but also realised if I had been after studying deer that day I would have had a five-mile climb-walk before the *stalking* even started. Something stirred at my feet – a fluffy buff-white Smoky Wainscot moth was winnowing its wings and scrabbling eagerly over the grasses after a female whose wings were more blue-grey and which was clinging with upraised body to an asphodel stalk. Nothing else mattered to him as my camera clicked and he quivered towards her, antennae vibrating, with all the amateurishness of the first timer, his white tail fanned out, trying to set it on hers. All that way I had walked and seen nothing but some deer in the distance and there, in the middle of nowhere, was the trembling ardour of young love!

It's all right for you, pal, I thought. I stumped back down the steep slopes, the camera pack heavy on my back as I negotiated an overgrown rockfall where tufts of heather concealed the gaps between massive boulders, and went home. Seven hard miles for a photo of a couple of moths!

I had known this would be a hard world to really come to know. I had yet to learn that these miles of wild open hills, so different in every season, had to be approached like mysterious, even desirable, gods, to whose secrets one would never gain access without patient perseverance, a certain stealth and cunning, and above all, love; that spring was the alluring nymphet, summer the mature but capricious goddess, autumn the crotchety god losing his prime, and winter a vicious old god of withered skin and frozen bones; and that while all their secrets were revealed but slowly they would be worth all the more for that.

Chapter 2

First Wildlife Adventures

After a week of early autumn storms, mid-October blessed Wildernesse with an eight-day Indian summer – the clouds rolled away, vanishing like battle battalions over the high horizons, and the sun blazed down with a fierce bright heat seldom felt during the misty hazes of the Highland summer. As it scorched through the window, made white paper dazzle the eyes, all desk work was impossible.

As I de-mossed and painted the cottage exterior, cut and carried fallen logs from the woods and hacked away the bracken to clear a vegetable patch and cultivate the wild grasses and flowers, the first wildlife happenings began. The woods were now as filled with bird songs as if it were spring – strident chipping wrens, the *pink pinks* of gaudy chaffinches, the silvery trills of robins seeking new territories after breeding. High in the conifers goldcrests zinged, and occasionally I caught glimpses of a great spotted woodpecker flashing his cancan colours of black, red and white among the trunks, or heard him drumming on a dead ash. From thickets came the chirring notes of tits and the slaty squeaks of tree creepers who worked their way in spirals up the trunks, searching for insects in the bark. Large clusters of blackberries had appeared on the brambles, and I decided not to clear these away but trim and cultivate them. Blackberries were as fine a fruit as any, after all. The hazel bushes were overloaded with nuts and I saw the red squirrel I had watched when I first camped out at Wildernesse the previous winter. He was leaping busily, tail flicking to help him balance, as he gathered the nuts and acorns from the oak trees.

That gave me the idea to collect nuts for my own winter supplies. Out I went with an old blanket and shook them on to it from the whippy branches, so as not to lose too many in the bracken. Even so I left more than I took for I could do without them better than the squirrel.

When I was tired from the outdoor jobs I just lay down and tanned myself in the sun. Then I could watch a great black and blue striped *aeshna* dragonfly which had emerged from its leafy hiding place during the storms. It hawked about, swooping with contemptuous ease to catch flies and gnats in the scoop formed by its massive jaws and thick black legs. Occasionally it came to investigate my face, hovering with rattling wings and treating me to the frightful primeval stare of a pair of bulging multi-lensed eyes that blazed like twin mirrors with a metallic sheen. Then it would spot some worthy prey, like a big buzzing blue-bottle, sweep up like a falcon at a finch and snatch it from the air.

Dragonflies have long been my favourite insects for it was they who helped start my love affair with nature. As a city boy of twelve on his first country holiday in Sussex, I had discovered a little wild pond in the woods, and above it floated these incredible creatures of darting fire. As they hawked over the reeds, snatching insects on the wing, they had 'minds' of their own, knew exactly what they were doing. Clearly, they were the dashing Red Barons of the insect world. Dragonflies are a miracle of evolution and have been hunting this way for 340 million years. Even the *aeshnas* we see today are Britain's oldest flying predators and were hawking the skies 150 million years ago, long before the first dinosaurs appeared.

Their eyes are among the most efficient in creation, and contain over 20,000 lenses. Few insects can see six feet in front of them, but if just one of a dragonfly's hexagonal lenses perceives a fast movement up to forty feet away, its lightning reflexes and acrobatic skills allow it to investigate immediately, make a capture – or flee. No wonder there has been no need to change its form in all that time.

It is sad that dragonflies are in decline, especially in the south of Britain, where many of their breeding ponds have been polluted, allowed to silt up or filled in for development purposes; where rivers have been 'improved', cleared of vegetation, or tainted with insecticides and herbicides; for these magnificent insects are immensely useful to man himself. Both as a larval nymph in the water and as a flying adult, one dragonfly kills thousands of mosquitoes, midges, flies, and wasps in its lifetime. When it hovers round farm animals it is not to sting them, as was once believed (it has no sting), but to catch pests like bot flies, whose larvae invade the animals' noses and stomachs, or green-bottles whose maggots feed on any injured flesh.

For days as I watched my hawker hunting about in systematic figure-of-eight beats I wondered if, like myself, he was an accidental solitary. Then, on 14 October, I saw him put the dragonfly's mating technique into action. If it left much to be desired from the romantic view it was stunningly effective. Male dragonflies don't fool around. Out of nowhere a browner female came winging round from behind the cottage. He shot upwards at great speed, hit into her with an audible clash, then seized her neck with the special anal claspers at the end of his 3½-inch body. Down they went, their wings buzzing noisily like a clockwork train left running in the grass. She seemed totally submissive to his headstrong desire and within a minute both flew back into the air, their wingbeats synchronising, and landed on a bracken fern where she curled her tail up under his body and mated. For several more minutes they flew round the area in tandem, then he let her go. I watched her fly to the loch's edge where she would lay her eggs on a water plant, while he flew, unsteadily now, to find a perch for the night.

He flew spasmodically for only two more days, his wings rattling as if with old age, his hunting days over, until on 17

October came the first small snowfall of winter and I saw him no more. He had perished, but not before completing his life's task.

By now the red deer were rutting high in the hills and occasionally, as I tried to finish the cottage repairs for the winter, I could hear the challenge roars of the stags. On 21 October I heard a loud roar near the cottage and went out with the camera to see a single old stag on the hill above. An old 'switchorn', with narrow inverted antlers holding only six tine points, stared back at me truculently as the camera clicked. Maybe he knew 21 October marked the end of the stag shooting season.

A poor specimen for my first stag shot at Wildernesse, I thought. I hope they aren't all as poor as that!

By mid-November the first winter gales were stripping the trees, heavier rains had turned burns into seething white veins in the hills, and the dull roar of the woodland waterfalls became the undertone to all sounds at Wildernesse. The gales and increasing cold turned the supply trips into more perilous teeth-gritting journeys, but at least some magnificent whooper swans, back from breeding in Iceland and northern Scandinavia, now flew over the boat, their long graceful necks extended, white wings whistling. Some mornings too a lone heron would be patiently standing to catch water creatures in the burn, unafraid of the red deer that might be grazing on the opposite bank.

The first heavy snow fell towards the end of the month, and after it had lain three days I felt sure that a hard winter lay ahead. But on a supply trip I was surprised to find the village completely without snow. It was only a few miles nearer the sea yet the climate was warmer. Again I wondered about the wisdom of my move, for by early December the mountains round my home had become white icebergs. My water system, a black polythene pipe from the burn, was often frozen up and I was reduced to carrying buckets of water for washing and cooking.

Small herds of red deer hinds and calves, driven down from the tops by the ice and snow, started to use the woods as dormitories. Early one morning I looked through the workshop window and saw a hind apparently dancing in the west wood. She seemed to be making a real sensual feast out of it, her feet moving to a kind of rhythm and her long neck sawing up and down to a regular beat. But when I got my fieldglass I saw she was not dancing in fact but using a small Scots pine as a rubbing post. First she rubbed her forehead on the rough bark, licked it with her long pink tongue, then with closed eyes rubbed the sides of her face gently, arching her neck like a bow so she could also work the bases of her ears against the tree. She paused, looked around, perceived no danger, then moved closer to the pine, and like a horse trying to shake off a bridle gave both sides of her neck a good working over. Next she switched herself round, backed up to the tree and with front feet firmly braced performed a cervine version of the 'shimmy' dance, rubbing her backside to and fro. Apparently this gave her great joy for her eyes closed, her face assumed a soppy smiling look and her lower jaw dropped open in a leer of pleasure.

Finally, her toilet completed, she stepped away, shook her coat so vigorously it seemed almost detached from her body, then followed the other deer up the slopes to graze on ling and heather. I went over later to find two of the smallest weedy Scots pines had had their trunks rubbed smooth up to a height of about five feet. Deer do this not only to dislodge pests like keds and ticks but for the insecticide qualities of the resin, which also gives the guard hairs of their coats a distinct gloss. They appear to enjoy the odours of its scent, too. They will rub on a favourite tree for two or three years until all the bark is off, the trunk wood becomes highly polished, and the tree dies. I looked up but both pines still had their full foliage of needles.

Two evenings later I came the closest I had ever been to a herd of wild red deer. I went out quietly at dusk, saw six hinds and two calves working downhill, and just had time to station myself under a large hazel bush. Wearing camouflaged jacket and bush hat, both washed in pine needle juice, I stood back amid the wands of new hazel growth and kept still. The deer slowly walked towards me, splitting up round the bush until the nearest was a mere ten yards away. An old hind with a grey face and torn ears with yellowish patches among the dark hairs, clearly scented something for she raised her head attentively.

The south-east wind was blowing what should have been my scent towards her but it was a dry day and with nothing danger-ous in sight her suspicions eased and she carried on grazing. They walked right past me and on into the front garden, and I could hear their cloven feet squelching in the marshier parts. It was too dark for photos but they were exciting moments. I even managed to get back to the cottage without startling them, moving slowly, an inch at a time, and freezing motionless at the first sign of suspicion. When I was back indoors, pleased with myself, they even stayed when I lit the paraffin lamp. I felt they were beginning to know that I was harmless. One night they appeared outside the window a few minutes after I had put some Beethoven on to my battery record player, as if attracted by the music. At least it didn't deter them.

On 19 December, after three days of sporadic blizzards, the snow lying a foot thick, I had a more extraordinary experience. I saw a hind walk across the garden to the west followed by (judg-ing from the width of its head) a stag calf. I watched from between the leafy branches of a rhododendron as the mother turned north to the first slopes of the hills. The youngster went slower, snipping off shoots and bramble leaves and nosing about under old bracken stalks. Then, as if trying to shelter from a new snow flurry, it walked behind an old ruined wall. Hoping for a close shot I approached quietly, reached the wall and peered round slowly. There, right before me, was its white-buff anal flash and two rear legs both trembling with the cold.

I don't know what came over me but with camera swinging from my neck I put my left hand on the wall and made a grab for the leg nearest to me, knowing the deer would leap and kick. It did, like a kangaroo, but I held on, moving upwards with it, and when it came down again I hauled it off its feet. Its other leg, the right, flailed away as I expected; if I had grabbed that one I would have had my face kicked in by the left. I hauled the hefty calf back and got my other arm round its neck as it snorted in my ear.

In those brief moments I saw it was in fine condition, spine well covered, mouth healthy, and I realised that if I had been starving, a man on the run, I could have killed it with a knife. Suddenly it let out a loud braying 'Bleah!' and I let it go. It just got to its feet and stood there, looking round at me, bewildered. I took a photo. The click startled it back to reality and it ran a couple of steps, then stopped to look at me again from behind a fallen larch. I took another shot. This time the click made it go, leaping the remnants of another old wall, and I was lucky enough to get a shot of the actual jump.

The calf 's mother was moving above in agitation. When the calf joined her they both stopped and looked back at me again, their expressions seeming to ask 'What the hell did you do that for?' I felt mighty pleased with myself at the time – to catch a wild deer by hand was surely a rare experience, and what was more I had the photos to prove it. Looking back, however, it was an unnecessary and foolish act, born of the hunting instinct latent in man, and I never attempted it again. Giving deer such frights in winter, especially if they are in a weakened condition, is neither sensible nor kind, for I have since found out they are susceptible to shock. Three hours later, however, both mother and calf were back, this time in the east wood.

As if that occurrence was not enough, I woke up some mornings later to see sleepily what looked like some brown sticks waving in front of the window. Then a big head with slit-pupilled eyes looked in. It was a stag, calmly observing me in bed! When he looked away I sneaked out, shivering in just a vest, and took a photo from the door. The trouble was that while red deer were now easier picture targets, the winter light was too poor for colour film shots.

I realised too that because the deer were now using the little woods for nightly shelter, their browsing, together with that of voles and occasional sheep, meant the trees could not regenerate naturally. Well, I would have to do something about that.

Now was the time of the winter solstice and the perpetual darkness in daytime, increased by the high hills and the nearby woods, was at first as depressing as it was bad for the eyes when I was working at the desk. It was far darker here than my home on Shona, under the wide open Atlantic skies, had ever been. I had to light the lamp around 3 p.m.

Two days later I boated back from a supply trip in a blinding blizzard. The slippery snow made it harder to pull the boat out. My mail brought depressing news – the novel which I had written three times over seven years had been turned down and for the fourteenth time and I felt it would never sell. The outline of my book about the Canadian wilds had also been rejected by a publisher who had first shown great interest, but finally didn't believe I had trekked alone in grizzly bear country. I sat for a long time in silence only relieved by the hissing of the paraffin lamp, trying to face the fact that I was now a failure, in every sense. No wife. No children. Hope receding, resignation setting in. And no human around for miles. No real friends left anyway. It seemed I had no chance either of even communicating a few experiences or thoughts by writing books. I had not only cut myself off for no purpose; I was also *being* cut off by forces beyond my control. It seemed almost as if society was exacting revenge for my daring to flaunt normal concepts of community and the need for companionship.

Next day, water pipe frozen up, I went to fill my buckets in the burn. There was a dead hind washed up on the rocks by the flow of the water. Her upper eye was missing. Her upper haunch had a gaping red wound where a fox had been tearing away the flesh which was in fresh condition. I took hold of the leg and heaved her out and found shotgun pellets also in the haunch. Only a fool would use a shotgun on a full-grown deer hind, so it was probably the work of a poacher who had shot as the hind was running away.

I back-tracked her twin slots through the snow up about two hundred yards into the hills and found a round depression between the tussocks where she had apparently rested. Around it were her droppings, made as she had lain there. How far she had walked while wounded I could not tell but somehow she had found the strength to stagger down to the burn and had died after stumbling into the icy water while trying to cross it. With the weather so rough now I didn't know when I would be able to get out again, so rather than waste the meat I sadly cut off the good haunch, skinned it and hung it in the woodshed. In fact I was hemmed in by gales for over two weeks and baked chunks of the haunch in tinfoil over the log fire.

Depressing news also came from the faint radio. Power strikes had plunged some cities and hospitals into darkness and it also seemed petrol rationing was imminent for coupons had just been issued. Even if I had had somewhere I wanted to go for Christmas and New Year, what with the petrol problem and the blizzards, the chances were I couldn't get south anyway. Walking back from a short trek one day I glimpsed the cottage from an unusual angle in the bare woods. It looked very isolated and lonely. Lonely house, lonely man, I thought. Good company for each other.

I did what I always do when the loneliness sets in – turned to the creatures of the wild for company. I was already putting out food scraps, and bright colours and cheerful calls flashed and sounded around them as little robins, tits and finches hopped on the window sill. If I couldn't improve my own lot I could at least improve theirs.

PART TWO

Chapter 3

A Bird Table Theatre

It was a robin which shamed, hustled, me into making the first bird table at Wildernesse. I was picking blackberries one November day when it came flitting through the bush on its knock-kneed legs, scolding with loud '*tick tick*' notes. 'It's all right. I'll leave enough for you lot,' I said. The robin flicked his wings towards me, bobbing indignantly as if to say 'Get out of it' or, being a Highland robin, 'Awae wi ye!'

At first I made a miserable effort, a square foot piece of wood on an old post by the kitchen window, and set a slice of bread on it each morning. Within a week the 'daily bread' was disappearing before dusk but I only had occasional glimpses of chaffinches, two great tits and a blue tit when I went into the kitchen to make lunch or a cup of tea. It took me two weeks to realise I had not been clever; while feeding the birds I had failed to provide free entertainment for myself. So I took a morning off from trekking and writing on 1 December and made a really picturesque avian canteen. I cut a slab from a thick red pine tree butt that had drifted on to a beach half a mile away and nailed a strip of wood along one side to stop the prevailing south-westerly winds from blowing away the food. On one edge, to give it a homely touch, I built a nest box of larch slabs, then set the whole edifice on a thick post a mere nine feet from my study window.

One mixed blessing of having no electricity was that I could not run a television set – and so could not succumb to the temptation of switching it on and slacking in my desk or outdoor work. But in following winters I lost many man-hours through watching this little wildlife theatre, so easy to make, cheap to run, and for which no licence fees, rates or taxes had to be paid. On this palace of varieties I have enjoyed as much comedy, farce, drama and, indeed, tragedy as from any television set.

The first character upon the winter stage, as befits the boldest or tamest of small British birds, was the robin. This jaunty clown who, in folklore, gained his red breast from being wounded when pulling spines from Christ's crown of thorns, or in another version, from being burned when taking a brand from the Devil's forge to bring man his first fire, arrived early one morning from the east wood. With long thin legs spread wide as if to make room for the pot belly, he landed on the spray of plum tree twigs I had nailed to the table for birds to perch on while eyeing the goodies below.

Soon he was joined by two smaller robins which at first I thought, wrongly, were both hens. Like many birds robins evolved 'unisex' long before man because the sexes look alike. While one of the smaller birds, clearly a hen, could feed without protest from either, the other two engaged in ding dong battles from the start. The plump cock would land by the food, glare with black eyes at the other cock, who came from the west wood area, then leap into the air with a quick wing flip, and down again. Then up shot the other and in a trice they were leaping up and down like two circus acrobats jumping on either end of a see-saw to send each other higher into the air. No beak or claw contact was made but suddenly the smaller one would turn tail and flee to a rhodo-dendron bush by the path.

Soon a squabble of four cock chaffinches arrived, making the table a whirl of colour with their blue heads, rosy chests, white wing stripes, green rumps and white-edged tails, as they piped noisy '*pink pink*'s and fussed busily about. Two landed on the twigs, one on the nest box lid while the other set about the bread. This bird didn't dip down for a nervous peck like the robins but went at it in true chaffinch style – dug a powerful wedgy beak in deep, twisted it from side to side, pulled out a lump, and was promptly 'jumped' by one of the others. I soon found that these colourful birds had a strict 'peck' order and they all looked slightly different. For instance, the biggest cock, with frowsted feathers and small silly eyes, was the lowest in the social scale, for he hopped sadly about emitting mournful peeps until all his mates had had turns at the food. Only then was he allowed in.

Suddenly – pandemonium! The top cock had arrived! He was sleeker, his eyes bulged and were set closer to the top of his head, and his white wing chevrons blazed with health. He landed, tail high, then lowered his head and swerved all over the table with speedy hops and open beak, until he had cleared all the others away. Then he took his time to peck and twist at the bread with a cocky dispassionate air. Because of his extra-thick wing chevrons I called him Corporal. He was the bossiest chaffinch I had ever known, which, when you know something about the male chauvinist world of chaffinches, is saying something.

When he had gone again, a robin and two of the other chaffinches returned. Now they were joined by a little blue tit. Arriving like a miniature d'Artagnan from *The Three Musketeers* in a pub full of bucolic boors, little black Chaplin moustache twitching with resolve, it soon proved it made up in wits what it lacked in size. The moment it saw a gap it leaped on the bread with both feet, took a few lightning pecks and was away like a tiny blue rocket to the top of his nest box, where it performed a brief pirouette of triumph, before the others knew what was happening.

Then out of the west wood conifers came the smallest of our tits – a coal tit flying as straight as a blunt-headed arrow. With plumage as ragged as an untrimmed beard and a big white patch at the back of its head, it looked like a little bald back street pawnbroker. It too was cunning, waiting on the wood strip until a chaffinch had finished eating, then diving in, holding the bread with both feet, taking quick little stabs, then heading back for the woods, tiny wings a blur. When two great tits arrived I was surprised to see that they were almost as big as the chaffinches, landed full in the middle of the bread, and took no nonsense from them when feeding. Whenever a chaffinch did go near a great tit it glared, swerving sideways slowly and menacingly with open beak, to warn it off. But when Corporal arrived and dived in, it straightaway flew away.

In the first few weeks I thought I had the sequence of domination worked out: the big robin could bully the other cock robin and the great tits but ignored blue or coal tits as if they were beneath contempt; the great tits could bully the blue tits but these could see off the coal tits; and the top chaffinch cock could put all the others to flight if he chose. How wrong I was I found out over the years, for the variations between the species, plus the individual differences within each species, were enormous and always vastly entertaining. Behaviour also altered according to the degree of hunger, the kind of food set out, and the harshness of the weather. My early conclusions soon appeared to be the notebook platitudes of the lazy naturalist.

When there was a heavy fall of snow the coal tits, robins and chaffinches dug down for the food with their feet but the coal tits always carried it to a twig and held it down with both feet close together before pecking. The great tits disdained snow digging and concentrated solely on the netting sock of nuts that dangled from the twigs. A coal tit was sometimes cheeky enough to share the sock but was ever wary, keeping carefully to the opposite side from the great tit. If it relaxed vigilance it soon received a hard stab or a wing-clout from the bigger bird. Robins had no appetite for nuts, as if knowing that they could not hope to hang on with their skinny legs and fat bellies. Only a chaffinch could husk an ear of barley or an oat in its beak. Perfectly still, only tongue and beak mandibles working, it turned the seed round and round, neatly shaved off the light skin and swallowed the kernel.

One sunny morning the chaffinches scattered, while a great tit and a coal tit crouched in instant freeze; then everything went quiet. Suddenly a male sparrowhawk, its golden eyes glaring, floated across the garden and landed on a horizontal hazel branch. His copper-barred chest feathers, fluffed out under each wing, were flamed into pinky orange hues by the sun and he looked around with that leisurely screwball ease, that head-on-gimbals look so characteristic of the sparrow-hawk, then he leaned off the branch and went on his way. The silence lasted a full minute more. I had kept and flown two sparrowhawks and knew that eerie silence quiet well. When the hawk flies in woodland glades the small birds cease to sing. I was sure this male's piercing eyes had seen the bird table and its flashing cluster of his natural prey, and had stored the image in his memory.

Next day I saw a demure hen chaffinch standing on the table, perplexed, staring at a great tit guzzling on the nut bag. As he swung to and fro she launched herself at the bag too. The tit flew away, but though the chaffinch hovered briefly it seemed she couldn't hold on to the bag and gave up. Two days later, after many hoverings, she worked out a way. Standing on the nearest twig, she extended her neck until she was half as long again as normal, then tweaked out a nut. I noticed when I put out a new slice of bread the great tits often bounced on it several times while in flight, like trampoline dancers, as if to test it *was* food and yet make a quick getaway if there was a trap. Once, after a snow fall, I saw a great tit take exception to a blue tit on the nut bag. It flew up and grabbed its smaller cousin with both feet then heaved it down into the snow. The little blue tit, fearless as a Highland terrier, emitted indignant '*tseeping*' squeaks, kicked back lustily and, seemingly unperturbed by the encounter, shot back to the bag first, extracted a nut and zipped away to eat it in peace in the willow bush near the cottage.

Often at dusk I would see a robin or blue tit on the table, desperately stoking itself up for the cold night ahead. The robin gulped down bread so fast he seemed almost to choke himself, the food stuck in his open beak. The dapper blue tit would arrive, take about fifteen pecks, flinging crumbs all over, and be away again – full up in some six seconds! Twice I saw the smaller cock robin, the one bullied by the larger, land nearby, startling the blue tit. When the robin made a direct pounce at the bread the blue tit stood its ground, stretching its titchy body as tall as it could, its beak up in the air, and the robin gave up and flew to the ground. How did the little tit know this robin could be stood up to yet it had to give way to the other?

One mid-December morning, a steady drizzle making its usual gentle brushing sound on my iron roof, I was woken in the pre-dawn gloom by loud '*cup cup*' notes. I looked out to see two cock blackbirds on the ground. They were approaching the bird table in fits and starts, nervous, wary and flicking their tails, yet were announcing their presence in this odd way. As soon as they saw my face they shot into the bushes with terrified shrieks, but were back next morning – and every morning when the weather was harshest. I didn't mind blackbirds being my cheap alarm clocks – but why in hell didn't they set themselves for an hour or two later!

Although I was glad to help feed the birds during the bleak wet and cold of winter, I did not want to create artificial stocks which would be deprived suddenly of the expected meals in such an isolated place if I had to leave for a week or two on a work trip. So I put out fairly small amounts, mainly bread, kitchen scraps, hazel nuts, chicken corn and a few cereals. When I did go away I left two whole loaves out, hoping they would last until I got back.

One evening the table had an unusual visitor. I looked out in the near dark, saw something on it and went out with a torch. It was a lovely little woodmouse with bulging black eyes and large ears. It sat up eating seeds like a squirrel, unafraid of the sudden light. There was drizzling rain in the air and occasionally the woodmouse scratched daintily, flicking the water from its ears and eyes like a little cat. The branch of the plum twigs I had nailed to the table stretched down to within a foot of the grass. Clearly scenting the food, the woodmouse had jumped that height, then shinned up the branch to the feast above. Suddenly, it shot away and down the branch like a little monkey and disappeared.

Next day the first real blizzards started, and when the radio said border roads were already closing, I dashed to the nearest town (a 44-mile drive away) for my Christmas shopping and more bird food. I returned in darkness, found I had stupidly forgotten my torch and spent three hours burbling the six miles home in the boat, desperately trying to count the gravel shores. Ensconced at my desk later, paraffin lamp hissing warmly, I looked out to see the woodmouse back on the table. I went out with some nuts and it seemed even less afraid, staying to watch as I set them down six inches away. How the tiny mite deduced I was friendly I don't know but it clearly had and, back indoors, I saw it gnawing away at the nuts.

At least I had one companion over yet another lonesome Christmas in the wilds. On Christmas Day he appeared again at dusk and this time seemed to respond to my little '*tch tch*' noises and actually let me stroke him with a finger after I had put more food down. I could understand how a convict in 'solitary' would look forward to a tame mouse coming each evening.

It was not until 10 January, a dark day of leaden rain, that I next saw a sparrowhawk. This time it was the larger female. She glided past the window like a vengeful shadow, as if summing up the situation for a time of real hunger. As I looked briefly into those staring yellow eyes I knew she had seen me sitting inside.

After the hail storms, rain and dull calms of January and early February, the morning of the 17th dawned in a sky of bright aluminium. The sun now cleared the tops of the mountains to the south and blazed down as if showing its determination to achieve again the high arcs of summer. It was so hot I sunbathed for two hours, and this was the day the small birds first began to sing. '*Tootsie, tootsie*' sang the blue tits from the birches, '*me too, me too*', came the thin pipings of the coal tits, and '*tee cher, tee cher*' sounded the metallic cries of the great tits.

The big cock robin tinkled his silver notes from the shade of a rhododendron and from high in the ash tree nearest the cottage Corporal chaffinch let out the first chipping notes of his cheerful song. If it was rare to hear all these birds give their first songs so early in the year, it was even more surprising that they had caught napping the bird that is normally the first herald of spring. Next morning, although it was dull again, he made up for it. From the highest sprig of the great spruce in the east wood I was woken early by the rich fluting of the Highlands champion avian piper, the mistle thrush.

The blue tits were the first to start actual courting. On 27 February I watched the male chasing the female and displaying all round her. He was flying like a lovely green-blue moth, wings shimmering rather than beating, so that he seemed to be surrounded by a mist of his own making. Sometimes he glided briefly with his primaries apart like a tiny buzzard. Twice the female landed on twigs, raising her crest and prostrating herself with open wings and fanned tail, but as soon as he came close she flew off again. By now the large robin seemed to have paired with the female and they fed amicably together on the table.

A few days later I saw the blue tits in the east wood, their wings a blur of emerald. The male landed on a loose larch twig which had become hooked up, balanced over a branch. Down swung the end. Off he flew and up it swung. Down he went on it again as if showing off to his lady love by see-sawing in this way. As the tall larches waved softly in the wind, moving in different directions like long river reeds stirred by water, I could hear his high-pitched calls '*ping ping, zee zee zee*' echoing through the forest.

Maybe he had some success for he grew really cocky and came to the table with optimistic bravado written all over his face, even prepared to take on a chaffinch. When Corporal grabbed a piece of bread on which the blue tit was feeding, the little tit, probably weighing less than a wren, refused to let go. He clung on as the hefty chaffinch dragged the bread around as if on a sleigh ride – but he did look fed up!

By this time the mistle thrush was ringing in every dawn; blackbirds piped with mellower querulous tones from amid the catkins on the hazel bushes and the cock chaffinches filled the woodlands with their bright tripping notes nearly all day. How varied were their little end phrases, ranging from '*Dear, cheerio*' and '*Switch it radio*' to one in the west wood who had apparently named himself '*Philip Spio*'. I swear I even heard one always ending with a perfect pronunciation of '*Shakespeare*'.

The odd thing was that all these different endings occurred in the same area as the season before, and when in the seventh spring I heard a '*Philip Spio*' from the west wood I was sure it was a son of the first. Chaffinches live no longer than five years in the wild and inherit their song patterns genetically, but individuals listen to other cocks when they are young and work out little variations of their own.

Chaffinches certainly do not adhere to any avian concept of women's lib – any cock could always see any hen off the food, and usually did. The poor hens had to leap in and grab what they could when the cocks weren't looking. Now the cocks fought at the drop of a crumb, and also for the attentions of the two duller-coloured hens. I was often startled at my desk by the loud whir and snap of wings – one would be at the food, another would appear and then they were shooting straight up into the air as if up an invisible drain pipe, a fluttering blur of pink, blue, black and white feathers, pecking and clawing at each other until one fled. Once, in a ground battle, I saw Corporal drag another cock's tail right over his head, and the loser flew to perch in a bramble bush looking most abashed. Well, what Highlander could retain his dignity when his kilt was thus disarrayed?

It was comical to watch the two top cocks pairing off with the hens for then they *did* allow the females to feed nearby. Every so often they would stop feeding to glare at them, as if only just able to control the urge to drive them off. The hens had their own peck order, though, the dominant one persecuting the lower as constantly as Corporal routed his rivals.

The robins also stepped up their little battles. Once I saw the smaller loser throw himself on to his back and try a little kung-fu, kicking out with his feet. A few days later I had another good laugh from them. The big cock drove his adversary off the table and pursued him right to the ground. There the smaller bird just hopped a few inches away, then burst into a few bars of his little liquid song, as if trying to mollify the other. The victor sang not at all but just flitted back to the grub.

I also found the big mistle thrushes were aggressive at this time of year. One morning I saw a male thrush chase another through the high larches with a more demure female tagging along behind, as if egging him on. When the other suitor had been driven right across the burn, back came the victor and, as if titillated by winning the encounter, quickly mounted her with a great flurry of wings. Even the coal tits could get nasty with each other. Once I saw this tiny bird land right on top of a rival and twist round suddenly as if trying to screw off its head with both feet!

When I acquired a dog in my second year at Wildernesse he moulted hair all through the spring and this became an added attraction to the birds. The chaffinches and great, blue and coal tits hopped about looking like little Colonel Blimps weighed down by heavy moustaches as they gathered up the soft warm hairs for their nests. To help them I pinned tufts to the table itself. I was surprised once to see a bright yellow-green siskin land and fill its beak with hairs too, then fly to where it was building a nest high in a silver fir.

On 25 April in my third year a pied wagtail landed on the table but not for food. It too gathered up hairs, walking about and flipping – not wagging – its long tail up and down. It undulated away over the east wood flying unusually high, for it kept well clear of the tree tops. Sometimes it came back with a mate, probably the cock, for its white parts were brighter. Not until 30 May did I solve the mystery of where the black and white tail bouncers were going. I found the hen sitting on seven pale bluish eggs with light brown freckles in her nest on a mossy bank on the nearest offshore islet. They had come over a quarter of a mile just for the dog hairs which now thickly lined the nest.

I was sure pied wagtails would never come to the table for food, for they eat insects and can catch flies on the wing almost as well as spotted flycatchers. But in late March in the sixth year a pair arrived as usual from their more southerly winter haunts and competed with the chaffinches to peck at bread crusts that had been pulled to the ground. The bossy chaffinches seemed put out by the unusual visitors, eyeing them warily instead of trying to drive them away. Pied wagtails are less common in the south since modern farming methods and pesticides have reduced their insect prey, and I wondered if this indicated a slow change in wagtails' diets. Of course it could have been due to the abnormal hunger after their migration and the dearth of insects at this time of year.

The first hint of tragedy in the bird's little world came in late April in the first spring when the male sparrowhawk came out of the west wood and made a sudden dive towards the table. Its sole occupant, one of the chaffinch cocks, flew off, not with its usual undulating flight but straight like a brilliant rocket, into the thick of a rhododendron bush. The hawk seemed unconcerned, as if not really trying, and flitted casually on into the east wood. Two days later I spied the female hawk flying idly across the garden, as if just eyeing the scattering chaffinches. While I didn't want them to persecute the bird table, I have special regard for sparrowhawks and hoped they were a pair that would nest in the woods.

One morning Corporal's mate attacked another hen on the table – and Corporal actually went to help her, pulling a beakful of feathers from the luckless female's back before she flew away. That was taking even chaffinch chauvinism a little too far.

Two days later bossy Corporal's reign came to an end. I was watching him chase off a little cock chaffinch which had an odd brambling-like crest on his head when there was a swift shadow, a clash of feathers, with some flying into the air, and the female sparrowhawk had hit him. As my face appeared at the window the hawk let go and Corporal was left staggering through the air, his right wing damaged. I rushed out and caught him as he fluttered to the ground. There were three spots of blood on the wing but as far as I could tell it was not broken. As I was in a hurry to boat out on a supply trip I hastily made him a 'cage' from a large cardboard box into which I put hay, twig perches and food, sewed some string netting along the front and set it on the kitchen table.

When I fed him later Corporal seemed in fair shape apart from his dragging wing, and was certainly not suffering from shock. Not once did his cocky air leave him. He ate seeds and bread avidly despite my nearness, and made rattling noises with his beak when he was thirsty for a tin lid of watered milk. Next day I made him a better cage of wood and wire netting and he '*pink pink*'ed away as if being indoors in a cage was quite usual for him.

Within three days his trailing wing had come up well but he could not extend it more than three quarters of its normal distance. For two more days he hopped about the kitchen like a canary, ate with gusto and even sang his chippy little songs. Next morning, able to fly fairly well, he began fluttering at the window. What should I do? Keep him as a pet, leading an artificial indoor life? Or let him go and take his chances? He seemed to have accepted me so readily that I felt sure he would return for food if he was in trouble. I opened the window wide, put food by his cage and went to do some writing. When I came back an hour later he had gone. I saw him that afternoon when he came back to the bird table, but now he could not fly up to it. Suddenly one of the other cocks he had formerly bullied attacked and he fled with fluttering flight to the rhododendron bush by the path. I felt sure he would realise next day where his bread was buttered.

A pile of grey-tipped feathers below the table and some more pink breast and three tail feathers by the path told the sad tale next morning. Corporal had been taken in the night by one of the owls who ranged the lochside woods. The code of the wild

– early death to the unfit, unwary or foolish, so they don't pass on their genes – had claimed him in one night. Yet there was an ironic justice to it: if Corporal had not been so busy pulverising one of his weaker brethren he would have seen the hawk coming.

By the end of April, with the daffodils in decline around the cottage but celandines and wood anemones blooming in the woodland glades, primroses lining the burns and bluebells starting to cover the lower slopes, the bird table became more neglected. Most of the birds were feeding more on insects and tree buds. The greedy chaffinches were the last to leave, mainly the unmated cocks, and by early May they sat singly in bushes in the woods, their aspiring early songs replaced now by a single mournful '*weep*' note. After a few days this turned to '*weep-ticket*.' They reminded me of sad hostesses at old time dime-a-dance joints who couldn't find partners to buy their tickets! Finally, they just went '*ticket*', as if imitating bus conductor clippers.

On 6 May I saw the blue tit male feed his mate with bread that the chaffinches had pulled to the ground. She waited above on the table twigs while he flew down, took a beakful, then flew up and presented it to her with a sweet endearing gesture. If that seemed a nice end to my first bird table season I was even more delighted when accidentally I found the robins' nest in the west wood on 12 May; the hen shot out from a mossy stump, exposing five pale, red-speckled eggs to view. A week later the nest's dark cavity was filled with five scrawny necks and great yellow-edged mouths, for all the eggs had hatched.

Generally it's not good to put out cereals, seeds and bread scraps in the rearing season as many small birds feed their young on insects, caterpillars and other live food which the young need in order to grow strong bones and feathers. In my woods and natural pasture area I found the birds had so much live food they only used the table for a snatched tidbit for themselves while searching for food for their progeny. I was surprised on 25 July by one of the pied wagtails, which was showing three fledged youngsters how to catch insects along the lochside, flying to the bird table and jerking back cereals. It was a rainy day and she was just taking a snack herself while giving her young all the live food she actually caught.

In late May of the second year a strange thing happened. BBC Television had asked to make a film about my way of life and we had gone for a day's filming on the sea island where I spent my first Highland years. Not only did the seals I used to know come crowding round the boat as if to perform for the cameras, but Gilbert, a wild herring gull with an injured wing whom I had nursed back to health, also turned up. What was more, he had obligingly dived for scraps from my hand which no wild gull would ever have done on that remote Atlantic shore.

Two days later, as if he had gone looking for me again, he turned up at Wildernesse, landing with a scratchy clang on the iron roof. I went out carefully because it sounded as if an eagle had landed – and there was avuncular old Gilbert, pristine white and glaring down at me and the goodies on the bird table with bright yellow eyes. I went back in and heard his webbed claws shuffling above; then there was a flurry of long white wings by the window, and he was beating away down to the lochside with a whole slice of bread in his beak.

A few minutes later he returned for a second helping. It was then the common gulls which had a nesting colony on a nearby islet caught sight of him and two 'sentries' came winging over to drive him from their territory. He dodged the smaller gulls easily but, as the trio passed above, one of the islet birds saw the food on the table, wheeled abruptly and also landed on the roof. Soon several common gulls came just before dusk each evening to compete with Gilbert for food. At a slice a time this was becoming costly! But the snow white feathers and graceful movements of the long-winged birds were a beautiful sight.

Each year Gilbert arrived in late May, coinciding with the first midges rising from the loch, shrieking over the cottage at dawn to make off with food. He never stayed long or visited for more than a few days. Maybe he was helping to feed his youngsters on a sea islet or cliff, and so had to get back. Often a few common gulls flew about offshore, waiting to harry him if he went too near their islet. When Gilbert's slices of bread broke in half and fell into the water, they dived on the pieces with yelping cries and bore their unearned booty away, gulping it down in mid-air.

By the fourth year Gilbert flew over with deep '*qwuck qwuck*'s, rather like a loud blackbird, and landed directly on to the table's nest box instead of clanging on the roof. After a quick look round he jumped to the table, guzzled down scraps for a few seconds, then made off. Once I rigged the camera with a long air release cable and was lying in bed at dawn when I heard the familiar calls. I reached for the bulb. Almost instantly he came out of the sky like a white ghost, landed on the table, and I took my best shot of him there. It's not often you can take wildlife shots lying in bed!

One year he came only once, on 8 May. A loud kerlank on the roof was immediately followed by another; he had brought his mate with him! They '*keeyah*'ed loudly, flew down with flashing wings and landed on the grass. Then he walked round her with bowing motions of his head, wings a quarter open, making his shoulders look burly, and mewed softly like a cat. He was still courting her and it seemed he had brought her all this way as a special treat outing, for a ten-mile flight is nothing to a herring gull. Then, as I took photos, up they went to the table and pounded the sun-dried bread with their beaks to break it up. With a piece each they opened their long wings and glided away.

It is not until late summer that the small birds begin to return to the table, and then only on wet days when insects are harder to find and searching for seeds is a wet business. I looked out on 26 July in the first season to see a group of the ubiquitous chaffinches at the food. This time it was a whole family – cock, hen and their three youngsters. The cock would not let the young or the hen near his piece but did tolerate them on the table. When the chicks with their new but more ragged plumage wanted to be fed they weaved to and fro with their beaks open, trembled their wings and made little chirruping sounds rather like sparrows. The hen worked hard to feed them but the cock only did so occasionally, almost as an afterthought between his own gulps, and only then if there was a chick very close to him.

In the seventh year, however, a cock brought four young to the table and worked hard to feed them. I never saw a hen with this family. Maybe she had been killed by one of the hawks. Another year I saw a cock chaffinch fly up from the table to feed a young meadow pipit which had perched momentarily on the vane of my windmill generator.

By early August some of the young cocks actually fought with their fathers but the older birds do not fight as fiercely as in the mating season – as if just *teaching* their offspring to take care of themselves.

On 2 September I was surprised to see a grey wagtail land on the plum twigs, its bright yellow breast gleaming in the sun between thundery showers. It took a brief look round, spurned the food, then zipped up into the air after a fly. A few minutes later a bullfinch made its only appearance on the twigs in five years. This rosy-breasted cock found nothing of interest either and went to the woods for tree seeds.

Unobtrusive hedge sparrows were rare visitors, confined to severe winters when insect prey was scarce, but one established itself on the table on 23 January of the second year. To my surprise it drove even the aggressive chaffinch cocks from the food but when a little blue tit, which chaffinches could usually dominate, flew in and stuck its foot stubbornly on the bread, the sparrow flew off. On one occasion the hedge sparrow allowed the new top cock chaffinch to approach, held the edge of the bread with one foot and take a few pecks. But the chaffinch felt uneasy at being so close to the stubborn dunnock, which was standing four square on the bread, and so chased another cock chaffinch off a smaller piece.

Other birds I have seen singly over the years on the table have included wren, whinchat, stonechat, tree pipit, willow warbler, blackcap, brambling, goldcrest and tree creeper. None of these fed there.

Reasonably common though the colourful goldfinch is in the south of England, it is rare in the north-west Highlands and does not breed there. Imagine how I cursed on 9 December in the third year when one landed on the twigs, looked dolefully at the bread (I had run out of varied seed) and quickly flew off. I went out to find it had a mate and they were feeding off knapweed and ragwort seeds. With the proper seeds I might have induced them to stay through the spring and to breed. Another rare and once-only visitor, on 28 March in the fifth year, was a yellowhammer. It was raining heavily and the chaffinches soon chased away the odd man out.

The rarest visitor of all came on 12 May in the fourth year. A colourful bird I had never seen before, with a long white moustache, blue-grey back and a striking orange-pink breast, landed briefly on the twigs. I reached for my bird book and its section on rare visitors, which revealed it to be a subalpine warbler. The few vagrants that reach Britain in spring from south-west France and Spain are usually only found on southern English coasts. To see one in the north-west Highlands was a great discovery.

Only when it flew off, without touching any food, did I realise I should have reached for my camera first and the book second.

Chapter 4

Drama among the Birds

Although the chaffinches were the commonest and most colourful of the bird table tribes, they were mere noisy squabblers, socialising plebs, in comparison with the great tits. These handsome birds, with their yellow-green chests divided by a dramatic blue-black stripe, their oval white cheeks and shiny sloe-black heads, were the swashbucklers.

After I hung a cord threaded with peanuts in their shells from the twigs, only the great tits learned to cope swiftly. One landed, looked as if reasoning it all out, then bent down like a ballet dancer, grabbed the cord with its beak, hauled it upwards, gripped it with one foot and pecked at the top nut. But the long nut-filled cord was too heavy to hold for long like that and it had to let go. The great tit looked down again, then stood on one leg, reached down and grabbed the cord as low as possible with the free foot, heaved it back, pinned it down on the twig with *both* feet and finally pecked out a nut.

As winter approached only the great tits came at dawn to the window sill, to pluck off any flimsy November moths from the walls. And if there weren't any, and the table was empty, they dinged on the window pane with their beaks for breakfast, as loud as trip hammers in the twilight silence. They were the only birds to feed on the table by the light of my lamp in the early dark.

I was surprised in the second winter to find great tits would actually eat carrion. On 3 December I dressed out a red deer hind found dying in the woods, and left the carcass forty yards from the window below a rhododendron bush, to test the carrion-scenting ability of my wildcats.* Two days later I saw five great tits and two robins on the carcass. While some of the tits were taking the parasitic deer keds that were hopelessly seeking a fresh host, others were definitely pulling off meat. The robins seemed only interested in the keds. Next morning there were seven great tits at work. As they zipped from the leafy twigs to the carcass and back, the bush shimmered with new greens and blues.

They also seemed to have the largest musical repertoire of any small British birds, and over the years I identified over thirty different calls, including the familiar '*tee cher*' and the bright '*pink pink*', softer than the chaffinch call, which sometimes ends with harsh '*chir chir*'s to denote alarm. Most interesting were a loud metallic '*quick-ink*' to keep in touch with a mate, a soft greeting '*cer cer cer-wee*' made just before a pair began nesting, and an odd '*nnf ter-dink*' which seemed to be an alarm call to an incubating hen. So strident are some of the calls I was not surprised to learn that the great tit has been called the 'bell bird'. The male's courtship display can be comical – an absurd posturing strut up and down a branch, with drooped quivering wings, tail up, and the feathers of his head and neck puffed out with yearning.

In the second year, spurning two perfectly-made tit boxes in the east wood, one pair perversely decided to build in the rickety box on the bird table itself. They popped in to examine it on 4 March; then on 23 April both birds began carrying beakfuls of moss from the oak trees into the hole. Sometimes the hen, with a slimmer chest stripe, came back with an empty beak and dived in anyway. She seemed to be just checking up that her mate was building correctly. In three days of dawn to dusk work they had the nest completed.

* The whole wildcat story was told in two of my previous books, *My Wilderness Wildcats,* and *Liane – A Cat from the Wild*

What a difference came over their behaviour – the bird table was now *theirs*. While his hen was inside preparing to lay, the male clutched the plum twig spray, legs wide apart, glaring at any feeding bird that came close. He stretched himself up with elongated neck and swivelled on his feet like a matador, his wings slightly out, menacing the chaffinches with his beak, looking angrily up with little black eyes if they flew overhead. By now the robins had built a nest in the west wood again, but the big cock still courted his hen. He lifted his head straight up, puffed out his very red neck and breast feathers, showing tiny black patches at their bases, and swayed slowly from side to side, half singing his little tinkling song to her as he did so. She, doubtless impregnated and about to lay, seemed to take little notice.

I could not keep close watch on the little nesting birds for a while as I was busy with outdoor work and treks, but on 19 May the male great tit was bringing grubs to the hen all day, working beyond dusk by the light of my lamp. Two days later the hen left the box at 10.30 a.m. and didn't return until after 2 p.m. With the sun blazing down she clearly knew it was like an incubator and that it was safe to leave her eggs for a few hours. Her belly feathers were all ruffled, unlike the straight black waistcoat stripe of her unsitting mate. Strangely, the male was still bringing dog hairs and moss to the hole as well as grubs for her. Maybe he wanted to be sure she was not only well fed but tucked up comfortably too.

By 29 May they had hungry young and both birds began working hard. Now they brought in green winter moth caterpillars from my oak trees, plus a few mottled umbers. Both these larvae can destroy the foliage of oaks and birches, so the tits were doing me a forestry service. I noticed that they watched me through the window: if they saw I was looking at them, they waited until I looked away before entering the hole. Sometimes the male gave a squeaky cheep when his beak was full of food, as if saying 'Ready or not – here I come!' I timed their arrivals in late afternoon:

between them they were bringing one caterpillar or beakful of insects every four minutes. They fed the young faster towards dusk, as if stoking them up for the night.

When the parents were away, a light scratch on the box induced the young to make a thin silvery tinkling noise. If the hen was brooding my tap produced a loud hiss from her, almost as loud as a wildcat's.

By 8 June four young had flown from the robins' nest and several sexton beetles had found it and were laying eggs in the corpse of one chick that had died. They had located the little carcass by scent. Six of the seven fledglings from a wren's nest in a ruined wall had also flown, but here another youngster had died. In this nest, of moss and deer hair, the black and orange-striped sextons had been joined by woodlice, who only briefly shrank in their antennae as the beetles clambered clumsily over them.

I fondly imagined that when the great tit young first ventured outside they would all line up on the table twigs by their box to be fed, and I would be able to get some good photos. From dawn to dusk on 15 June the parents were working at great speed to feed the chicks on caterpillars. Next day they were still at it as I trekked up to the high peaks to try photographing red deer calves. To my chagrin, when I returned, the entire great tit family had flown the nest box and were all neatly lined up on the branch of a dying ash tree behind the cottage. The light was too poor for any photos. The hen looked a little tatty after all the hard work, but the cock was as immaculate as ever, chirring at me to keep my distance.

Great tits used the oak tree and bird table boxes over later years but a family in the table box came to grief in the fourth year. At the time I was trying to tame a young female wildcat (which I called Liane) and so far, in her first year of life, she had seldom shown interest in the birds.

On 12 May I awoke to see the wildcat actually on the bird table. I chased her off and hurried out: the entrance hole was covered with tiny green feathers and the box was empty but for the remnants of the nest. She had clearly reached in and scooped the sitting female out with her claws. For several hours the bereaved male chirred loudly from the twigs as if trying to call back his mate but I doubt he felt the kind of emotional sadness known to humans for he soon gave up.

Indeed, over the years it became clear it was every bird for itself on the table and there was no sentimentality, even among the same species. At breakfast one morning I heard a loud thump on the window – a female of a new great tit pair had for some reason flown straight at the pane. I rushed out but found her dying. Wondering what her mate would do I laid her body on its back on the table and went indoors. Soon the male landed on the plum twigs but merely gazed briefly at her lifeless form before hopping down to peck at the food as lustily as ever.

At least in the season when the wildcat took the female, the box under a big branch in the nearest oak tree produced a family of five young; and the nest in the table box was not wasted entirely. Two days later a pair of cheeky blue tits arrived, each perched on the edge of the hole, looked carefully inside to make sure there was no occupant, then went in to steal beakfuls of moss and dog and deer hairs for their own nest in the west wood.

The nerve of these 4½-inch mites amazed me. One April morning in another year a kestrel landed in the lochside ash trees, looked round briefly, then carried on into the west wood. Immediately a blue tit which wanted it off its territory, flew up, looped close under it and darted back. The kestrel turned, tried to pursue but the tit was already dodging through thick conifer twigs where the open-field hunting falcon could not manoeuvre. What guts! It was in and out before the kestrel knew what was happening.

One early May I was checking through a long wood a mile away when a hooded crow slipped off her eggs in an oak tree, swerved down through the trees so I couldn't see her against the skyline, dodging as skilfully as a woodcock, and vanished. Hoodies build deep nests in high leafy places where it is hard to see into them. I was just trying to peer down into this one from a high rocky outcrop when along came a blue tit, helped itself to a bundle of hairs and small feathers from the nest's lining and flew off.

Blue tits reveal extraordinary intelligence in adverse circumstances. One cold mid-February day I found one in the kitchen, and caught it, brought it into the study and set food in various spots, hoping if I stayed quietly at my desk it might become tamer. At first it flailed at the window but soon realised it could not escape that way and flew to the bookshelf to watch me in silence. Then it saw the food on the mantelpiece. Over it flew, stoked up with crumbs, then flew back to the window. Instead of beating its wings again it hopped up and down, probing for a hole with its beak, realised it *still* couldn't get out, felt hungry and flew back for some more food. When it was tired it hid behind some books for a nap. Then it renewed escape attempts, so I let it go.

On 3 July another blue tit, maybe the same one, came into the cottage. I caught it in my hand where it made no attempt to struggle. Knowing it preferred live food in summer I put some mince on to my other hand, held it near, and let it go. To my surprise, instead of flying off, it just hopped on to the hand with the mince and started feeding. I then put it into a cage of small mesh wire netting and set it outside to see if its mate would come to feed it, try to rescue it or show some concern. Its mate didn't get the chance. The tit perched quietly, examined every inch of the netting with its cunning little black eyes, then made a single dive and escaped. It had located one little spot, unseen by me, where a strand of wire had parted and through that double-sized hole, pulling with its feet and hunching its wings, it squeezed itself and was away.

Birds were by no means the only performers on the stage of my little bird table theatre. Several times in winter a field vole shinned up the pole and squeezed through a hole in the centre of the pine slab to spend several minutes at a time chewing at the bread and cereals. In the seventh winter I could not understand how so much food was disappearing overnight – until I found a family of five woodmice had taken over the nest box and were living in it. A cunning bunch they were for they always left a 'sentry' peeping out from the nest box hole while the other four guzzled away at the food. I was also amazed at their monkey-like agility when climbing and shooting about the twigs in the dark. The biggest surprise of all came in the eighth summer when I looked out of the window to see a huge squirrel-like face with bulging black eyes sticking out of the nest box hole. Surely the profile was too narrow for a squirrel, and it certainly wasn't one of the two red squirrels in the woods for its fur was a grey-brown colour.

Suddenly, feeling the coast was clear, the head shot out of the hole, followed by a powerful muscular greyish body. While the animal clung on to the inside of the hole with its rear feet, its jaws grabbed a whole slice of bread and jerked it backwards. The sides of the bread jammed on the edges of the hole, the slice (except for one large bite) fell back on to the table, then the big head again came out, eyes glaring, to see if it was safe to have another try.

It was a rat! A big brown rat, the first I had ever seen at Wildernesse. I set up the camera and managed to take a few pictures of it, despite the fading light. I have no aversion to rats as I know them to be fastidious animals, as clean as any creature in the world if they can live away from sewage and get food away from dumps and tips. This one looked as clean as my best linen, but I also knew a lone rat was liable to be a pregnant female look-ing for a new place to start a colony. I let it finish the slice before sneaking out with an empty washing-up liquid container filled with water. As I treated it to a cold shower squirted through the hole I also suggested in loud but reasonably polite invective that it really ought to find another home. There was a pause, a slight explosion as the back of the rickety box was forced outwards and the last I saw of this new competitor for the birds' food was the rat legging fast for the west wood, its long ropy tail thumping the grass on each side of its fleeing body.

Perhaps the most poignant little adventure on the bird table, for it had overtones of courage, occurred one early April morning. I was at my desk, colours flashing outside the window, when there was a swish, a loud snap, and I turned to see one of the cock chaffinches fluttering to the ground – and the female sparrowhawk wheeling round in a circle with beak agape as if to make another dive at the chaffinch. I threw open the window to see more and she veered off into the west wood. I rushed out but the chaffinch, flopping on the ground as if in shock, flew into the dark interior of a bush and disappeared. I hoped it had only been dealt a glancing blow.

Next day the chaffinch was indeed back but its left leg was broken and dangled loosely by a sinew from the main joint. It had probably spent the whole day and night in the bush and was now hungry. It stood on the bread with one good foot and pulled off crumbs, holding balance by spreading its wings out on the table. I now recognised the bird as the second dominant cock of that year, one I had called Pete because he would arrive raising and dropping his head feathers, twirling on his feet like a concert singer acknowledging applause, and calling '*pete pete*', slightly different from the usual '*pink pink*' of the other cocks.

I didn't see him for the next two days and wondered if he had perished, but after heavy snow blizzards on 8 April, I saw him on the table again. All the cocky aggression which had allowed him to hold second position among the seven cocks that year had gone, and all the others now bullied him. Once the top cock, dozing on a twig in a sunny moment, seemed to wake up for the sole purpose of chasing him away. But Pete learned to be cunning, to watch from a distance and to sneak in when the table was empty. As I watched him stoutly fighting to live amid the whirling snow, I thought how flimsy was our belief in human 'superiority' over such animal life. That night the radio announced the deaths of some hikers during the blizzards in the Cairngorms; men who were clothed and equipped were dying. How long would a man last in such conditions with a broken leg? Yet the tiny chaffinch was making it.

Two dawns later he came peeping breathlessly on the window sill for me to put food out, his leg still dangling uselessly. There seemed to be no infection but he looked in worse shape. When any of the other cocks came near he went into a begging posture, whether for mercy or in hope of food I knew not, but he rippled his wings like a fledgling. Regression to childhood behaviour when in dire extremity is not confined to humans. Once the dominant cock hauled him right off his one good foot by his wing. Later two of the others chased him into the bush by the path. There is no compassion in the animal world, only maternal, and occasionally paternal, care – and even that is short-lived.

Next day little Pete spent most of the time on the window sill, showing no alarm when I put my hand out with food. He pecked at his dangling leg and foot, trying to pull them off. There is no national health service in the lives of the creatures of the wild. When the other cocks tried to take his food I banged on the window. They went but, as if he knew why I was doing it, Pete stayed. That afternoon, the snow having changed to a constant drizzle, I made a box shelter for him on the sill. He took to it instantly, pecking at the crusts, but at night he flew to roost in the bushes. I so admired his courage.

On 15 April he stood on the sill and finally managed to peck off the useless leg. He now learned to balance better, using his tail as a second leg when landing or tugging at bread, and using little wing flips so he didn't fall over as he hopped about. He also started fighting back at the other cocks, although he lost all the ground battles. A fine spell followed and I only saw him occasionally. When he used his box shelter on 25 April, I saw he had grown new flesh over his stump. All the chaffinches were now battling and trying to pair up, but as there were only three hens to seven cocks it was clear I was not going to be the only unmated male around!

Three days later Pete was still being picked on by the other cocks but held his own by bluffing, opening his beak wide and hitting out with his wings. He was sticking it out, gallantly fighting to live despite his handicap. By the end of the first week in May, with many leaf buds open, weeds growing and insects flying, the chaffinches and other birds were more independent and the table was often deserted, yet I noticed Pete chased after insects more than the others.

For a long time I didn't see him at all, then in early June he landed on the table, sprawled a little, grabbed half a crust of bread and flew off with it, labouring in flight, towards the west wood. I felt sure he had just taken the food to a safe sheltered place to eat.

As the days passed and I never once caught sight of him, although some of the unmated cocks still came to the table now and then, I thought he had probably died. But in late June a hen chaffinch landed on the twigs and had only been there a few seconds when a cock also arrived, plucked off a lump of bread, flew up and popped it into her open beak. As he shivered his wings for balance I was surprised to see it was one-legged Pete. It seemed impossible that out of seven cocks and three hens he could have been one of the cocks lucky enough to have won a mate.

I searched sporadically for a nest but came up against the irritating obstacle I faced every year with the chaffinches: here was I feeding them through the winters, the commonest birds at the table, yet their nests were such masterpieces of disguise I could hardly ever find them! Tucked into a branch fork or built so cleverly on a high branch of mosses and similar coloured lichens, they looked just like small swellings of the normal bark and were almost impossible to locate. What was more, only the greeny-brown hens did the incubating, as if their flashy mates' colours could have been seen too easily. Certainly I never found Pete's nest.

A week after I had seen him feed the hen, the dominant cock came to the table with his mate and three young, all males. The hen was feeding one when Pete arrived, darted at all the others, ate some bread, then lost an aerial battle with the father, who had always beaten him before. Two days later I saw a hen and two chicks on the table, then Pete arrived. At first I thought they were the dominant cock's family again but as the two fledglings fluttered their wings and weaved to and fro chirruping for food, Pete took a few pecks, looked at the nearest chick, a big cock, then popped some food into his beak. This was *his* family! To my further surprise, when two of the other chaffinch cocks landed the big chick went straight at them, chasing them away, then returned to act like a baby again, soliciting more food from his one-legged father.

When one of the chicks begged for food two mornings later, Pete looked at it, hopped over, pecked it on the beak lightly, then went on feeding. The youngster got the message and began to peck at the bread itself. During the summer weeks, though they all fed mostly off insects in the trees, the chicks also vied with their parents for food on the table on rainy days, and on 2 August I saw Pete actually let his biggest chick drive him off some.

A few days later I noticed he no longer fed his own young and one morning he flew after and caught a crane fly. A chick weaved to and fro and opened its beak hopefully. Pete seemed about to pop it into the chick's mouth but then sat back on his tail, clipped away with his mandibles as first one wing fell off, then the other, then he swallowed it, looking very pleased with himself.

That was the last time I saw him. He did not return that winter or any other winter. If the sparrowhawks, an owl or just the harshness of trying to survive in his wild world with just one leg had finished him off, at least he had won his courageous fight and had left issue, and his mark, upon the chaffinch world. He had not lived his little life in vain.

Chapter 5

A Pup Called Moobli

In return for helping them in winter the birds on the table were an entertaining substitute for the give and take associations among my own kind. Even so I was soon conscious that they had their own societies, mates, and that my role was merely one of provider and sidelines spectator. At that time I had lived alone in wild places for most of seven years. I was now up a roadless loch, over six miles from my nearest neighbour, and while I was living this way to study and write about the inspiring natural world I still felt pangs of loneliness. The company of the little birds was not enough.

Often my thought returned to my old companion Booto, the wild stray dog which had adopted me when I lived alone in a cliff top cabin in Canada. If he had been too old to sentence to long confinement in British quarantine kennels when I had left Canada four years earlier, he was clearly much too ancient now. I still heard about him from friends, and knew that in old age he had become slow and stiff-legged. I would visit him before long but I was now convinced I ought to get a new dog.

Booto had been part Alsatian and I wanted a big *doggy* dog who would be a companion and help track wildlife, so I thought in terms of this breed. Eventually I contacted a lady breeder in Sussex, renowned for her large but amiable Alsatians. She replied to all my queries, said that all her pups had superb temperaments and none had ever chased sheep and that their natural instinct was to protect their owners and property. She could reserve the pick of the dog pups in her next litter and have him ready to leave for Scotland in early January.

I immediately sent off the deposit and, when the time came, boated out and drove south against snow blizzards to collect him.

When I first saw the eleven-week-old pup I was finally to call Moobli walking in his breeder's courtyard, I was instantly dubious. He was as fat as a piglet, his vast paws seemed out of all proportion to his size and unfolded clumsily on the ground as he moved. His great ears flopped like a pair of wings and to me, he looked more like a bloodhound pup than an Alsatian. The breeder again showed me his pedigree – seventeen champions in just four generations. When I picked him up he licked my ear – at least he was affectionate. As we drove away in the Land Rover he cowered a little from the racket of trucks whizzing past on the motorways, then slept fitfully with his head between his great paws on my lap.

Four hours after dark we limped into a fir wood south of Glasgow. There I cooked him a bigger supper than my own, as was set out in the breeder's diet sheet, the first glance at which had me worrying about my meagre bank account. He snored raucously on the front seat all night but as he never tried to invade my bed in the rear it seemed a small price to pay.

As I completed final shopping in the village at the end of my home loch, the pup attracted attention from the locals. One asked if he were a cross between a St Bernard and a dachschund!

The waters were rough when we reached my pine wood parking spot six miles from home, but I managed to manhandle the 500 lb boat down over pine branches and hold it steady while I loaded it with gear and put Moobli on the deck. To my surprise he showed no agitation – just braced his thick forelegs and semi-webbed paws against the surging movements and stared at me solemnly along the length of his long muzzle. We had only gone a few hundred yards into the stormy following waves when the engine gave out. I took the fuel hose apart as we wallowed and blew it through, petrol spurting over the deck, then reassembled it. Suddenly Moobli yiped and lifted his paws up and down; the film of petrol was stinging his toes. I lifted him on to a dry sack and felt his body quivering. Once I held him up to see the land that was now to be his home but he was frightened then, whining for the safety of the semi-cabin. The vastness of the towering mountains scared him and he shivered for the whole stormy journey. I was afraid it had all overtaxed his young strength.

When we landed I put the pup out first but he must have felt too alone for he climbed up a rock, took a leap for the boat, missed, and fell into three feet of water. He swam out, shook himself, and when I came ashore leaped about wagging his tail. He soon proved he was a tough little chap and seemed happy to be at Wildernesse. I showed him the garden and woods. He soon learned not to entangle himself in my legs, traipsed gamely through marsh and clambered over fallen trunks. He loved to play, shoving an old shoe about with his nose, or chasing a broom as I swung it round in circles. Pup or not he looked quite frightening as he chased it with an intent stare, his small pup teeth clashing noisily. He even retrieved thrown sticks and balls.

At the end of the first week, my initial euphoria faded. He suddenly started behaving to my inexperienced eyes like a juvenile delinquent. He took to doing his toilets in the kitchen, though each time I showed him his crime, said 'Badog!' and shoved him outside. I tried leaving the rear door open at nights but each morning there it was again, slap in the middle of the carpet. He began delving into the waste box, throwing peelings all over the floor, chewed holes in the carpet despite all his playthings, and ripped the plum twigs off the bird table.

Although he had accepted it at first, he now kicked up an awful fuss when left to sleep alone in the kitchen. Once, when he hauled a deer haunch off the top of the calor gas fridge, I slapped his rump and he turned on me with a growl and would have bitten if I hadn't swiftly withdrawn my hand.

I tried being soft, congratulating him with a food tidbit when he did things right, coddling him at night before leaving him alone to sleep. Now he hated being picked up and whined with peevish grunts if I put my head near him when he was on his bed. If he was like this now, what would he be like when powerful and mature? He became recalcitrant in other ways too, and I began to feel this particular dog and I were not right for each other. One night, feeling foolish, I wrote to his breeder admitting it was probably my fault, but maybe I would be better off with a bitch or a less independent breed of dog.

Oddly, the day I posted the letter things began to improve. We came back in a terrible storm, the boat pranced like a rodeo horse but Moobli, feet braced against the bags of supplies, never took his eyes off my face. As we landed safely in moonlight, he came to lick my hand as if he thought I had saved his life. One evening I was cooking supper when I heard a violent choking noise: Moobli had half a biro pen stuck across the rear roof of his mouth. I reached in carefully as he yiped and removed it. Instantly there came a look of gratitude over his face.

I invented new games which helped to unify us. One was kitchen football, which we played if it was raining. We bounced and chased after a plastic fish net float, but Moobli usually got the better of the exchanges, four feet being better than two.

In due course I received a justifiably indignant letter from his breeder, pointing out my mistakes and saying she would be glad to have 'super pup' back if that was what I wanted. I was glad to reply that Moobli was now settling down well.

Naturally he was interested in any animal he saw on the hills but any move towards sheep or deer elicited a quick 'Na!' and a slap on the rump. In later life he saved the lives of several sheep by scenting them after they had been caught up on brambles or had got stuck on sheer ledges where they would have starved to death.

For exercise, I taught him to jump from the boat when we were close to shore coming back from supply trips, and to run beside it as I made for home. Within a further month he had lost his puppy look and was sleeking down into a healthy young Alsatian. I began to take him on longer treks. We had just passed two lochans on our first high trek when he came up to me, lifting his nose, stepping high, and scenting the air. I realised there must be something ahead. We rounded a knoll and there, less than forty yards below, was a herd of red deer hinds, yearlings and calves. Having been warned in this way, I already had my camera out of my pack, and got several photos though the light was poor. Already Moobli's nose was proving useful.

As the weeks passed and he grew in size and strength, his memory for commands increased and we became closer, largely through games. I varied the original puppy routines. I found that by throwing two sticks we had the ideal exercise: he chased after one hurled fifty yards away, picked it up, ran back with it, dropped it at my feet, then tore off in the opposite direction after the other, all without slowing his running speed. Just how fast he was becoming I found out when he was a year old.

It had not taken him long to learn that sheep were taboo and he soon lost interest in them on treks, but his fascination with deer remained strong. One day, after he had already covered a mile on his exercise run along the boggy, tussocky shore as I boated home, he startled a group of hinds grazing behind a ridge. He got their scent, saw them and, now I was not with him, the temptation to chase proved too strong. Well, he would not catch up to fleet full-grown deer anyway, I thought, the boat being too far away to call him off. To my astonishment, as they ascended the precipitous slope, Moobli bounded upwards powerfully like a cougar and caught up to them within two hundred yards. The rearmost hind stopped, saw that he was immature, and lashed out with a forefoot. Moobli then turned and amicably trotted down to the shore again. I pulled in the boat, reminded him he had done wrong, and gave him a sharp slap.

Training a large and powerful male Alsatian is not easy but I noticed that, provided I was *fair*, there was no tendency to rebellion for its own sake, unlike a human male in adolescence. I constantly encouraged his scenting abilities so he would be a valuable ally in my wildlife studies. When he sniffed out fox scats or their scented urine on mossy rocks or tussocks, I urged him to 'Track the foxy,' and he would scent along narrow trails, where they had jumped on to old logs, and find more of their marking posts. Over the years I was able to work out many territorial movements of foxes with his help.

When I bred and released rare Scottish wildcats back to the wild in a three-year operation, I was only able to keep up with their rangings by Moobli's ability to track them down and put them up trees so that I could check their condition. He was able to locate by scent many badger setts and fox dens which, because I was either passing too high above or too far below in rough terrain, I would not otherwise have found. On his runs along shore he located scores of deer carcasses, two dead owls, two otter holts, numerous spraints, many nests of ground nesting birds, pine marten droppings, even the pellets of birds of prey. He became expert at scenting deer long before we crossed a ridge or rounded knolls or craggy inland cliffs, alerting me to ready my telephoto lens in time. Once he even located a golden eagle eyrie by scenting the carrion on the nest as we walked along a cliff top.

Such feats, and many others, came later. All I knew after the first few months was that he was turning out better than I had ever thought possible.

PART THREE

Chapter 6

The Loch in Winter

The loch is like nature's eye set in the earth between the mountains, reflecting the moods of the sky, the changing colours of the hills, and in the lush season the leafy trees fringe its shores like rows of lashes. Like an eye, it responds sensitively to every breeze, every arrowing zephyr that ruffles its surface, and to summer gales that send its waters into surging billows of green and blue that end with turgid crashings of milky foam upon the white banks of gravel. In winter storms its sombre fathoms are lashed into black troughs, and pounding waves as glassy white as the frost or snow that lie upon the land smash against the jagged outcrops of granite. But it is on a calm blue day in spring that it is at its most beautiful, when the bitter lash of winter has departed and the migratory birds return to bestow upon its broad molten bosom their brilliant flashing colours, the sight of their love displays, the music of their voices – the living beauty for which the lonely human dweller on its shores has waited for so long.

It is then the water carnival begins, with a dipper landing with a slight plop near the boat bay, swimming buoyantly in the mini-billows, diving its plump white-breasted body below the surface for food. Sometimes it clings to a submerged rock and preens itself. Each time it dislodges dust specks from its plumage with screw-like motions of its head, along comes a wavelet to slap the dirt away. Once I was lucky enough to see it walking underwater, holding on to stones with its feet and angling its back so the flow of the burn pressed its body down as it searched upstream for

water larvae and nymphs. A pair of gaudy shelducks appear by
Heron Island, their startling greeny-black, chestnut and white
plumage flaunting the 'law' that female ducks are always more
dowdy than the drakes, for only the knob at the top of the drake's
bright red beak distinguishes him from his partner.

By now the first common gulls have returned from winter by
the sea to roost on the nearest islet where they will form a breeding colony; already the dawn is filled with their white flashing
wings and their shrieking cries as they dispute tiny territories. As
the fish life stirs from the winter depths of the loch, and the
hibernating frogs come out to look for mates, a lone heron floats
in each dawn like a blue-grey wraith, to fish the shore or try its
luck along the burn. Now too are heard the eerie drawn-out
mewing wails, like those of a large lost cat when heard from a
distance, of the great black-throated divers who have returned
from wintering on the coasts further south. No sound is more
evocative of Highland wildness than these mournful calls, and as
these rare birds preen and wash, throwing themselves sideways
on the misty surface, their creamy bellies send glimmering reflecting ripples across the loch. Now and then a few red-breasted
mergansers, back from the salt water estuaries, swim by in pre-
mating groups, calling to each other with harsh prolonged
'*kawyaw*' squalls. The untidy dark green double head crests of
the males make them look like country yokels with greased hair,
all spruced up for a trip to the seaside. Along the shore sandpipers are staking their territories, and as we walk down to our
boat they clearly feel we are the trespassers, for away they go
piping shrilly '*swee de tit, swee de tit*' with flickering gliding
flight, to land on a stone not far away, bob their tails and then
pipe us aboard.

Just as these spring players arrive, so do winter characters
depart. The white-fronted geese, with their glossy dark barred
chest plumage, leave the mossy area at the west end of the loch
and forage all the shores for the best new grazings before their
long flight back to the Arctic tundra where they nest. At dawn

and dusk long skeins travel westwards down the loch, already in shallow V formation, calling softly with haunting bugling cries as if assuring each other to be of good cheer and have courage for the great 700-mile flight ahead, before they vanish into the horizon. Now too we lose the white angels, the magnificent whooper swans which have used the loch and its wetter pastures as a winter dormitory. One morning I see fourteen sliding gracefully along, bright yellow beaks parting to emit muted '*woop woop*' notes, as if already courting. By early April the last of them have left to nest in Iceland, northern Europe and Russia, small groups passing so low over the boat I hear their wings singing like harps.

Yet my sadness is momentary for I know the springtime splendours that lie ahead – the dippers, the gulls, the ducks and mergansers, the sandpipers nesting along the shore, the curlews and lapwings in the marshy meadows, and the beautiful black-throated divers perhaps breeding again on the islets. I know that the first salmon will soon be running, driving through the river shallows at night to summer in the loch before spawning up the burns in late autumn, and that the sea trout will not be far behind. I am sure that the myriads of elvers will be shimmering in from their 3,000-mile journey from the Sargasso Sea, and that now I may again see an otter's whiskered submarine head cleaving the water as it returns from the rivers to feast upon the stirring fish life amid the green caverns of the newly growing weeds in the deeper waters.

For all its beauty at this time of year, however, I came to learn, long before I had Moobli, that the loch was my master, that it controlled my physical destiny, that on the long boat journeys to fetch supplies it cradled my very life on its swelling deep, and that if I failed to treat it with care and respect it would fashion my end.

During the hail and rain storms of the first November I learned the difference between handling a small boat on a long narrow fresh water loch and on the open seas. Here on the loch the winds piled waves into short troughs so that even three-foot-high ones smashed into the boat, being steeper and more dangerous than ten-or twenty-foot waves spread out like liquid downlands over the broad swell of the ocean. They were hard to kick out against and I had to have the engine primed with petrol and choked before launching, for there was only time for two pulls on the starter rope before being swept back. Then off I would go, the boat banging down into the deep troughs after breasting each wave. I had to sit on the floor up front against the worst gales, so my weight helped stop the prow from flying up and back and the boat turning turtle.

On 13 November I learned the best way to handle the hail squalls, at least when coming back with the prevailing winds. I waited in the boat below my Land Rover in the woods until the last stones of one squall were falling, then set off in a trough of relative calm. By timing my speed I could keep just behind the squall that dragged its grey skirts up the loch before me, yet still stay ahead of the squall that threatened me from behind. Thus I could cover the whole journey in choppy but not dangerous water. Occasionally I even got the last load up into the cottage before blackness and pelting hail descended once more.

By December, when gales were stronger, I looked for these 'pockets of peace', where the wind died off ahead, like a ship-wrecked mariner looks for a rescue boat. On the outward jour-neys, against the winds, I dashed to make the slight lee of this bay or that before the next deluge struck. More often I just had to slow down, keep the bow to the waves and take the pounding, unable to see more than a few yards because of the density of the hail or rain.

On 13 December a radio in the local store announced a Force 11 gale 'imminent' in the area. As I would at least be going *with* it, and it was getting dark, I took the risk. I had only covered half a mile when it hit the loch. Dark clouds rode up behind us on blue and violet palfreys, stinging the face with shrieking air. Deep in the troughs the prop laboured away but slowed down as the following wave pressed hard against the boat while she was held up by the back of the giant wave in front. Slowly she surged up it, then hastily I had to throttle down as she hurtled with a roar into the next trough, seeming about to go to the bottom. But, thank heaven, she came out of it each time, her bow rising in a shower of spray, water pouring sideways off the semi-cabin and back into the boiling surface. Whirlpools of wind hit the loch like hammers on ice, skirling up spindrifts so that six-foot water spouts shot into the air with disconcerting suddenness.

As we ploughed and lurched to my own shore I knew it would be hopeless to try to get the boat on to the trolley – I would have to beach it on the grassy bank now covered with water. I only had seconds to play with. Keeping a keen eye out for submerged rocks, I headed in, at the last moment turning the prow to port into the waves, briefly gunned the engine to swing the boat right round, slung out a full calor gas bottle and the heaviest bag of supplies, and leaped out myself. Then I grabbed the stern and using the force of the waves, now being safely split by the bow, hauled the boat backwards on to the grass. There, with the waves smashing the bow up and down, I unloaded in seconds, seized the bow rope as one would the bridle of a bucking horse, and, on a wave, hauled it in too. By running from end to end, I managed to heave the 500 lb weight up bit by bit until the whole hull was beyond reach of the pounding loch. Athletes could die here, I thought, as I plodded up in the near-dark with my precious mail, little knowing such trips, and worse, would have to be repeated many times in the years ahead.

While it may be thought I could lay in stocks and sit them out, such stormy periods often lasted for many days on end. Besides, it was not supplies I had to go out for regularly but post. In the first six Highland years, before my first wildlife book was published, I earned most of my living by journalism. Editors won't wait weeks for answers: deadlines have to be met.

In the first months it was hard not to believe that the elements themselves were trying to wreck my wilderness living. One stormy night I was woken by two loud thuds and I went out to find a dead alder had blown down, missing my boat by two feet. The boat too had been blown right off its trolley – though luckily it had landed on the grass.

Foot by foot, from early November, the loch rose higher in the rains, until finally I had to remove the winch from the shore and set it on a post high in the garden, which itself had to be braced with cable to a stake. Some winters the loch rose so high I had to winch the boat right up to between my gateposts – a full twelve feet above its usual summer level. And I had fondly imagined that living by a fresh water loch meant I no longer had to cope with tides.

Another winter problem was the short Highland day. Plan though I might, sometimes I had to boat home in the dark. Before I had Moobli, in the first December, I wanted to shop for items unobtainable locally and went out early in snow blizzards to spend the day in the town a forty-four mile drive to the east. On the way back the snow was so thick it jammed the windscreen wipers, and when I reached the wood where the boat lay at anchor, it was totally dark, the loch showing up as a vague blue streak. Stupidly I had forgotten my torch. I carried everything down, striking my lighter flint to see by its brief flashes the rocks and pine roots I had to negotiate, then set off into the ghostly blackness.

It was not too windy but I had to blink my eyes constantly to clear them, trying to make out the line of the shore. I daren't go fast and soon my gloved hands were freezing. I sat in the stern, one hand on the throttle, the other tucked behind my knee for warmth, the engine set in the 'Release' position in case we hit a rock, the blizzard blowing snow into my eyes. It was as black as pitch, just the tiniest glow coming from the water; like going into a pot of ink, as the engine burbled the boat along. I made mistakes thinking I was nearer home than I was, only the shingle shores showing up as faint lines. Paddling in to each one, the engine off, I struck my lighter – not here, not here, and on again. I had only the set of the mountains to go by, barely discernible against the dark sky, and I cursed myself for not knowing all the rises and falls in the hills from the pine wood to home.

At last I thought I recognised the outlined tops of trees along a burn where I could also hear water tumbling, but coming closer it was clearly not mine. I was afraid I would smash into the gulls' islet just off my own land. Or was I already past it? Maybe I would end up at the village nine miles on and would then have to drag all the way back, looking for the right shore.

The wind increased and I felt I was in real trouble as the boat began to buck. I recalled how in the old days of travelling as a journalist I felt put out if blizzards ever forced me to park overnight and seek a hotel. Here, in the snow and gales, on an icy loch with rocky shores, it wasn't inconvenience but death that awaited the failure. Finally I saw a black object with a furry top ahead. Was it my islet? I kept in close, praying no rocks would hit the prop or hole the boat. Then the white side of my old sea boat showed up as a dim glow, and the outlines of the west wood firs against the sky. I had never felt so glad to get back to nothing but isolation and work.

My problems were far from over. I landed with the usual sideways twist but as I unloaded gear the wind started to increase. In about ten minutes waves would really be ramming into the shore. I would try to get the boat on to the trolley in the normal way. I raced as fast as I could through the dark to fetch my torch, wasting time finding the right door keys, then set it between two rocks so I could halfway see what I was doing. I carried one of the movable runways to overlay the bank and put the trolley on it. Then I pushed the boat out into the waves stern first so I could haul it forwards again on to the trolley. Every time I got it halfway on, the heightening waves filled the lowered stern, making it too heavy to haul further. I had to let it down again, bail the bucking bronco out with a bucket, and try again.

When I finally got it fully on to the trolley, tied on, and was winching it up as it see-sawed madly with loud splashings in the raging water, one of the trolley wheels came off! In the end, unable to find the lost wheel-nut in the foam, I had to lever the quarter-filled boat up the runway with four-inch by two-inch lumber struts, jamming each down with rocks before setting the next, until I had it high enough for safety. Body perspiring but hands and feet still frozen, I eventually got all the gear into the cottage. It was nearly midnight: getting the boat out of the water had taken two and a half hours.

Next morning I found the nut and the wheel-holding bar had been dragged out by the receding waves and, amid cold sleet, I had to wade out and bend for them in three feet of icy water before I could repair the trolley and once more get the boat winched up.

It was becoming ever clearer to me why Wildernesse had known no permanent year-round human occupation since 1912! But this was just part of the price one had to pay in the all-seasons wilderness life, one that tourists and summer fishermen were lucky to escape.

Dangerous as some of these trips were, I could still extract some humour from the farcical aspects. Once, after four sunny days, I headed up the loch under a strangely leaden sky, dark to the south with streaks of violet and indigo in it, like the entrance to mythical Vulcan's forge. I knew that sky all too well but I had handled many storms and, needing to post urgent mail, and make phone calls, I chanced it. As I turned into the last dog-leg a real gale was howling from the south-east through the mountain gap and the waves chased me to my anchorage below the pine wood. It was hard to drop the anchor in the right spot and leap out fast for the boat pitched madly and the wind was so strong I had to shield my eyes from the spray.

Even when I had hauled it laboriously out to the anchor, taking advantage of the slack rope each time the bow went down, the waves were coming in so fast the boat still looked as if it was speeding along. Afraid it might drag the anchor and smash on the rocky beach, I intended to rush through my business and hurry back. In the village the wind was negligible: assuming it had died down round the boat too, I hung around until people I wanted to phone were free from conferences or had returned from long lunches.

When I got back to the anchorage the wind had reached near hurricane force. The anchor had dragged and the slewing stern was perilously near some submerged rocks. There was no way I could set off until the gales died down a little; meanwhile I would have to reset the anchor further out. I pushed off, leaped in, opened the little trapdoor in the prow of the semi-cabin and heaved on the bow rope. I pulled so hard the bow was forced down into the ploughing waves, which broke over me as if I was on the foredeck of a submarine – smash, smash, smash. The anchor would not budge. It had caught up under some tree roots and was trapped.

All right, I thought, as I leaped to shore again, the anchor can't drag any further but if the raging gales kept up the ropes would chafe through and the boat would still be smashed up. As it was getting dusk, all hope of returning home that day had gone. I would have to camp out overnight. I raced to the village for some tinned foods as I was famished, having had no lunch and not feeling like peeling and cooking vegetables on the little slow stove in the truck.

When I got back the wind had veered and the boat was now nudging some rocks. I dashed down in the semi-dark, looped the loop on the slippery pine roots, but managed to land mostly on both hands, so preventing an injured back. As it was, a shaft of pain shot up my left arm as my hand twisted on a rock, and it lost feeling for days. But saving the boat was my only concern. I would have to cut it loose and haul it out. I collected fallen branches and laid them along the water line, then leaped in the boat as it pranced crazily and grabbed what I thought were the correct two ends of one of the three ropes that went through the ring on the anchor. I tied these to a plastic canister which would act as a float to show me later where the anchor lay, then cut through the other two ropes to free the boat. Instantly the boat slewed sideways and headed for the rocks. I leaped out, held it off briefly as I kicked some branches below it, and with superhuman efforts hauled it over them, end after end, until it was out of reach of the raging loch. As I looked out, gratified that I had at least cheated the waves, the plastic float drifted up on shore. I had cut and tied the wrong rope ends, and now it seemed I had lost my precious anchor!

The strong winds were also buffeting the truck. As the trees creaked loudly under the strain of the gales, there was a real danger one might fall on it overnight. I drove back along the track and pulled into a treeless spot in a small meadow. I took a few swigs of wine and read some boring mail while I waited for my tin of stew to warm up on the camp stove. Then the gas bottle ran out, and I ended the day miserably spooning down the cold claggy stew before shivering through the night under the one damp blanket I had in the truck.

Next morning, after pouring rain, I found I was stuck in a bog. I threw branches under the wheels but the truck only dug itself up to the axles. I hiked to the home of my neighbour, farmer Donald (we lived six miles apart), whose tractor I had once helped haul out of some mire. Donald came with the tractor, his cows bellowing when they saw no sileage trailer attached. We broke first my rope, then his rope, and he roared off again to return with some doubled fence wire. This time the truck came out with a sucking squelch, and after a welcome cup of tea in the farmhouse I drove back to the boat, thankful that the wind had now dropped.

I heaved the boat back into the water and paddled about in it until I located the anchor by sight under eight feet of water. I couldn't leave it there as I was unable to function without it. Leaving on only my vest and a pair of socks, a trick I had learned in Canada when fording cold rocky-bottomed rivers, I took off the rest of my clothes and waded out. Then, taking a deep breath and gasping curses against the cold, I dived, twisted the anchor free with frantic movements and, by kicking against the rocks below, managed to scramble ashore with it. My hands and feet had gone blue in the icy water and, despite the socks, I had gashed an instep. I dried myself a bit with the damp truck blanket, got back into my clothes (all but the vest) and somewhat the wiser for the whole ludicrous experience boated with chattering teeth the long way home.

If such winter events were hard for a lone human, life was even harder for wild creatures. On one snowy drive I saw two buzzards and two tawny owls perched separately along the roadsides, hoping perhaps for a traffic-crushed rabbit or vole now the thick snow made their normal hunting almost impossible. Herds of red deer hinds, down from the snowy hills, sheltered in the loch shore woodlands. They ventured out at daybreak to scrape with their forefeet for grazing where the snow, melted a little by the relative warmth from the loch waters, lay thinnest. Their only pleasure at this time of year was to be found at their rubbing posts.

It was during my first Christmas at Wildernesse, when I had been feeling miserable, I realised that compared with the animals I was lucky! Lonely Christmas Day or not, I had books, a store of paraffin, log fires, records, letters to read, even a few cards and a parcel, and a roof that didn't leak. Also, I was a free man. I gave up sentimental wallowing in nostalgia, packed away photos of world travels, of old pals and of girls I maybe should have married, put camera in my pack and went for a trek along the shore, starting midway up the burn.

Suddenly there was a flurry a few yards ahead, a short 'zink' cry, and away shot a dipper between the banks, leaving rippling circles in a calm eddy. With its short wings beating so fast from its chocolate brown football of a body they were as blurred as a humming bird's, it sped upstream, made a last swift zig-zag and vanished into a small waterfall. I kept my eye on the spot as I got nearer, then was astonished to see what looked like a fish with wings below the surface of a bright pool. It was the dipper swimming two feet down, its wings thrusting it forward powerfully like paddles. Because of the air bubbles trapped in its plumage it glistened like a giant blob of mercury moving along.

I cut back down to the loch and had only gone half a mile when the sun suddenly blazed through a gap in the clouds, lighting clumps of tussock grasses and rushes into yellow and orange flames and dazzling the crusty snowcaps on top of dark patches of heather. Not until they moved did I realise some of the heather clumps were in fact grazing hinds, so well did their grey-brown winter coats blend with the broken background. The wind was east-sou'easterly, slightly favouring the deer, but being dry was not the best for carrying scent. I slid and elbowed my way through the crisp snow and took photos at forty yards. Then I stood up, expecting the deer to flee. Instead they just stood and looked back at me! Shooting them with a gun would have been child's play. As I swerved back to the shore, they just trotted up the hill, then stopped to watch me pass by, looking oddly like bottles on stilts when seen from the front. My hands grew cold on the long metal lens and I put the camera back in the pack. I soon regretted it, for just as I set off again there was a loud '*woof woof woof*' of great wings above me – a huge eagle, almost black against the sky and with an oddly grey mantled head, had soared silently over a ridge to my right, had just caught sight of me and was now speeding away.

As I walked through the bogs among the shore-side alders, placing my wellingtoned feet carefully between the rocks and tussocks, I was startled by a disturbance in the rippling waters. A dipper shot to the surface and without a pause rose into the air with the ease of a miniature Polaris missile, to land, bobbing its stumpy tail, on a rock 200 yards away. I now realised what an extraordinary seven-inch marvel Britain's smallest diving bird really is. I could think of no other water-bird that could duplicate such a feat. On the way back a male sparrowhawk flitted ahead with shallow wing beats, like a floating cuckoo, landed in a tree to take a good look at me, then was off again before I could get the camera out. Hell! I carried the camera and long lens in freezing hands all the way home but saw nothing else of interest, save a fat woodpigeon feasting on holly berries, well obscured by the prickly leaves. But at the end of the six-mile snow trek I wouldn't have changed places with a living soul.

Although I found a few otter spraints on shore-side rocks, it was not until Moobli came that I (or rather he) finally located the otter's home. I was boating home on 9 December in his second winter, while Moobli ran along the shore, when suddenly he wheeled smartly, put nose to ground, then set off again at a fast trot. To my surprise I saw an otter shoot up a good thirty yards ahead, gallop sinuously along the rough sward above the beach and dive between large rocks on the bank. From far out, I saw Moobli track it true, but as he stood with wagging tail outside the hole, I heard two high-pitched screeches. I called him off, not wanting the otter to be upset further. Moobli turned round to look at me, the otter dived between his legs in a blur of grey-brown, Moobli gave a loud yipe, and the otter vanished into the water. When Moobli returned as I was hauling out the boat he was limping badly and I saw an inch-long gash across the knee joint of his rear left leg. The otter had apparently delivered a slashing bite on its way through. The wound had to be stitched up later by a vet in the inland town.

On another Christmas Day, rather than mope with whisky indoors, I rowed to the big river to the east and walked with Moobli between the leafless alders. Out in the flow the body of an old stag, probably washed off its feet in the recent floods, had caught up in a tangle of dead branches against some half-submerged trees, the huge body rising and falling, twisting grotesquely in the current.

We headed on through more alders and willows until we came to a golden pool in an ox-bow bend where abruptly I hissed Moobli to stay back. Out in the pool, no more than two and a half feet deep, I saw what looked like two thick torpedo-shaped sticks moving. I peered through the twigs with my fieldglass. They were not sticks but large salmon! Almost side by side, they were lazily twisting their tails just hard enough to maintain their position against the current.

Suddenly one of them made a swift dart sideways and I saw two more salmon shoot away wildly as the first drove towards them in what looked like a semi-circular attack. These three fish looked darker than the larger one, which stayed where it was. As the attacking fish showed its sides briefly in the weak sunlight, I saw a distinct lower jaw hook and that it had patches of blacks and browns on its flanks, as well as a few small areas of rusty red. This was obviously a cock fish in full 'tartan' breeding uniform and he had actively driven the other two males away from the female.

The big hen fish went over on her side, then made a deep swerving scooping dive, down and up again, flicking her tail up so sharply from the golden gravel the suction alone sent a handful of it slightly downstream. She did it again, the same violent scooping movement, then settled back into the slight hollow she had created, as if feeling with her pelvic or anal fins the depth of the 'redd' she had made.

I was sure she was cutting out a redd in which to lay her eggs, and that soon the cock would swim close to her and release his milt to fertilise them. Then she would cover them up again with the gravel scooped from her next redd, which she would make upstream, so that they wouldn't get washed away. Unfortunately I was unable to witness this next stage, for both fish slowly turned and drifted more into the centre of the river where the water was deeper and I could no longer see them. I was surprised at this rare sight, not only because it was the first time I had actually *seen* salmon breeding, but also because it was occurring in late December, for I had always believed salmon in the west Highlands mated in November at the latest.

One day I met the new ghillie who ran the seasonal salmon fishing for the estate owners on one side of the river that flowed from the loch. He was wondering whether to dredge out much of the river weed and also to dig out some of the gravel, so making artificial pools in which to 'trap' the salmon and make it easier for them to scoop their redds.

'I can't understand why there aren't bigger salmon here,' he said. 'It's a good wide river with easy access from the sea, no falls to speak of, up to a big loch that is still fairly undisturbed. The average weight of the catch this year was seven pounds three ounces, yet there have been some twenty-pounders in recent years. Many years ago there were forty-four-pounders caught in this river and I want them back!'

I said I was no fish expert but that it seemed to me the extensive net fishing by the Danes, Greenlanders and Faroese off the salmon's main sea feeding grounds near Greenland and Norway, where many of the bigger salmon were caught, must have made a difference. The fish then faced netting by the Scottish licencees along the coasts of the sea lochs, the trawl nets pulled by boats, plus rod fishing in the rivers and an increase in fishermen on the actual loch in recent years.

'Maybe if you want more fish it might be an idea to *stock* the loch and river, not just plan on taking more out!' I said. As for digging out pools, it might be a good idea, but I was not sure about dredging large areas of weed.

I had seen that these weedbeds were a main feeding ground for gulls, the winter visiting geese, whooper swans and ducks like the goldeneye. Removing a lot of weed might interfere with the natural rhythms of the river, allow flood waters to blast out more eggs and fry, remove the breeding areas for much water life which fed so many birds and useful mosquito-killing dragonflies. Salmon also needed weed to rub off the sea lice as they swam to their breeding grounds. It would be better to make a full study of the river over several years first.

It was in late February, when Moobli and I made a twelve-mile return trek along the loch, that we finally found the otter's holt – in an unusual way. It was a day of extraordinary contrasts, the sort when spring inserts a hot cloudless day here and there, telling winter it is time to go – the sky is dark, rain hisses down and all the landscape is suffused with a grey-green colour, then the clouds roll back, all is brilliant blue and the sun is baking hot. We plodded over bogs, hummocks, small cliffs and rocky burn ravines, through twiggy scrub willow and bog myrtle bushes. Now and again the first rising fish splashed on the mirror of the loch. We were half a mile into the long wood when Moobli got a strong scent and began to hurry. I hissed him back, hoping to see ahead what he was after, trying to get my camera out of its pack as we walked. Then I saw a dog-like animal standing on the shore, head lowered – a fox, taking a drink. I whipped off the lens cap, wavered the telephoto lens about trying to locate the fox through it, but when I peeked to take a second sight to pinpoint the fox again, I was just in time to see it swerve up from the shore, its tail following like a dark wedge-shaped banner, and dash into the bracken caves by a fallen birch.

We hastened over there, but not a second glimpse did we get. It must have sidled along the far side of the trunk, out of sight, and up the gorge of a small burn, de-scenting its feet in many rock pools so Moobli could not track it. The first fox I'd actually *seen* at Wildernesse and I'd missed it.

Coming back along the shore line, hoping to find otter traces, we had passed the stump of a huge broken alder when I saw a thin trail of small grey feathers leading into some large mossy rocks. Moobli sniffed intently at a small gap and I found the remains of a woodpigeon stuffed into a natural cavity, though its skull had been left outside. At first, because some of the feathers had been neatly clipped, I thought it had been cached there by the fox. But surely a fox would have removed it from near the deer path, which was used by wildcats and badgers too. Then I saw more feathers leading out the far side and down almost to the water. Perhaps it had been found dead or killed by the otter.

We trekked on homewards until I realised we must be near to where Moobli had been slashed by the otter in December. Suddenly he got another strong scent, tracked it back and forth over more boulders, then paused above a tiny gap, tail wagging excitedly. I went to the shore edge of the small cliff and peered over. Much moss had been worn away from a large flat rock which stood at the entrance of a series of rocky tunnels. As I climbed down there was a strong smell of decaying fish.

Moobli scented on and moved to my right. Then I saw a brown form move in the dark hollow of the first tunnel entrance. It was big, moving sideways, rather like a giant vole. It turned and looked straight at me – an otter, barely five feet away. And of course, my camera was in my backpack and still had the impossible telephoto lens on it!

Was the otter cornered there? If so it might attack. But it just bared its teeth with a loud hiss as Moobli worked down to me and headed to my rear left. Then it backed away, turned and went on down out of my sight. I fumbled with the pack cords to change the lens on my camera as I dodged to my right so I could see into the next tunnel. I was only just in time to see the otter sneak through the adjoining crevice, come out on to rocks wetted by gently slapping waves, then slide with total silence into the water and make a foot-kicking dive to safety.

Heart beating fast, I replaced the telephoto with the standard lens, knowing I had no chance unless it came up within a few yards. But it didn't show itself again at all. Knowing an otter can range seven miles in a single night, that a dog otter at this time of year does not live in one holt but has several dens and lying-out places in its territory, I left the area alone for several days. When I went back I found its lie-out place, just past a small birch clump on the far side of the holt. A deep curved bowl of moss on a flat rock directly above the water was all pressed flat and still wet from where the otter had probably lain after dawn fishing. I spent many hours downwind of the place in the next few weeks but not another sight of the otter did I have. The trouble is that an otter can swim almost a quarter of a mile underwater and can spy on its holt from that distance while remaining unobserved by the human eye.

Photographing otters in the wild has, for me, resulted in sheer exasperation. Every wildlife photographer has a 'bête noire', a creature which he wants to capture with his lens but which always eludes him. Otters have been mine. I had just launched the boat for a supply trip on 15 March when I saw the otter on a flat rock less than 500 yards from the cottage. It held a fish between its tubby wet webbed paws and looked at me with its mouth still open after chewing. I had no camera with me. Just then Moobli appeared over the high ridge below the west wood – and plop went the otter. It did not reappear, but Moobli scented out one of the animal's spraints or droppings on a large rock. Otters leave their spraints for three main reasons – to help meet others of their kind in heat, for a female with young to warn other otters to stay out of her preferred hunting area (up to seven miles of shore or river) and for males to inform other males of their status, in size and maturity, so that actual fighting is rare as the juniors tend to avoid the seniors.

After we found the otter holt I made sure Moobli did not go near it again on his shore-side runs though, luckily, he never actually looked for animal dens unless I was with him. Mostly, he just ran!

The otter seems to waddle awkwardly on land but it can run faster than a man for a mile if it has to. One winter when a film of fresh water running off the land froze on top of the salt water at the head of a sea loch, I saw an otter running and sliding for sheer fun. It ran three or four jumps forward, then slid about twenty-five feet on its stomach with almost no loss of speed, then it ran a few more steps and slid again. Perhaps it could even save energy this way!

Otters are absent from many parts of Britain where they were once common, with strongholds in south-west England, East Anglia and Wales, but most of all in western Scotland where they have increasingly used the wilder sea as well as fresh water lochs and less disturbed rivers. Owners of fish farms dislike otters, for it is costly to make the tanks completely otter-proof. But otters kill thousands of eels (a favourite food, preferred to salmonids) as well as pike, frogs, mice, voles and occasional rabbits. When the otter was given protection in England and Wales in 1978, it seemed to me ludicrous that Scotland was not also included – and I said so in articles, and on radio and television. The Highlands are Britain's main reservoir of this dwindling beautiful animal – and if we don't protect such areas they cease to be reservoirs. At the time the otter was thus first protected, a local Highland landowner misguidedly invited a Warwickshire otter hunt to try over his estate. To my surprise the hunters were not received well by other locals. One man, who netted salmon for part of his living, was so fond of an otter he often saw that he threatened to shoot the first hound that went over his land. The hunt retired otterless – and hasn't been back in the years since. A two-year survey (1977–79) by the Vincent Wildlife Trust showed otters were not as plentiful even in Scotland as had been thought. Conservationists and wildlife bodies kept pressing until finally the otter was given protection in Scotland too, in the Wildlife and Countryside Act of 1981. Not before time.

One April morning the following year Moobli scented out a large spraint on the fishing rock, with a much smaller one right beside it. I kept sporadic watch for several months, sneaking my lens over a ridge forty yards away, hoping for photos of a mother otter and cub. Finally, one late-summer morning, I secured some pictures of a large otter swimming and climbing out with a fish. But my otter humiliation was not yet over. After a disastrous trip up the loch on 14 September, when my small boat went down in a gale, I dashed up the beach with my diaries and photo-sheets to keep them dry when the briefcase burst open and several sheets blew away. I recovered all but one sheet of my second litter of wildcat kittens, and of course lost the one containing the otters.

One afternoon I went to an old riverside holt nine miles away where I had known an otter to rear cubs in my first Highland years, before I took up photography. Although I found newish droppings along the bank, a new ditch had been dug alongside the holt which was under a rockpile. There were no prints in the mud and the den was filled with brown leaves, obviously not being used.

It was not until my eighth year that I managed to take – and retain – some photos of an otter in the wild. I was driving along the shore of a sea loch when I glimpsed an odd triangular shape in the water, rather like the dorsal fin of a porpoise. It was travelling parallel to the waves but heading in to the shore. Then I saw some seaweed moving beside a rock. I grabbed the camera from my pack, leaped out, crawled through a ditch, getting my boots and trousers soaked, and wriggled on my belly over big stones to a square foot-high boulder. I slid the long lens over carefully and focused it just as the big otter climbed out of the water on to the weed-covered rock with a small crab in its open jaws. I took three shots of it eating the crab before it treated the weeds as a sort of bed, rolling over on its back and wriggling its body and thick tail from side to side. After that, it turned the right way up and used the weeds as a whisker and mouth cleaner, wiping first one side of its jaws and head and then the other, rather like a human uses a towel.

Suddenly a car roared past noisily on the road above and the otter dived back into the sea and vanished. This time, however, the photos came out quite well, and I managed not to lose them.

Chapter 7

The Loch Gull Colony

They come winging up the loch like tiny white angels towards the end of March to take over once again their nesting islet near my home, so it is long before the twin tones of the cuckoo are heard in the glens that I know spring is on its way. What extraordinary biological clocks the common gulls must possess for they always arrive within a few days of 20 March – never, after a mild winter, earlier than the 16th, or after a long harsh one with snow still on the ground, later than the 24th.

Usually a single gull arrives first, flies round the rocky islet with its crown of little pine trees as if to make sure, lands hesitantly then stands there, the personification of doubt and loneliness. Or it may land on the water nearby, looking at the sky expectantly and calling out a few times before taking off again. Within days more arrive and then each dawn is filled with ringing cries of '*akeeya, akeeya, kew kew kew*' as they squabble over future little territories before they all fly off in a flotilla to feed from the weeds of the river bed or the seashore nine miles away. Replenished after an hour or two, they fly back to the islet, then stand about the rest of the day, chivvying each other before flying off for another feed-up.

The colony swells to thirty or forty birds before they begin to pair off. How they do this is as fascinating as it is amusing. Just before dusk about a dozen gulls form a straight line in the water, about a foot apart. The gull at one end paddles swiftly along the line like an army chief inspecting a rank of soldiers, turns and tries to fit itself next to the gull of its choice. If it is accepted, nothing happens and these two gulls remain conspicuously close to each other. Then the next gull also paddles along the line and fits itself next to, say, gull number five. If this gull rejects the suitor, it rises into the air with loud shrieks as if repulsing its overture, then floats back down again into the same place. The discomfited suitor then swims jerkily on down the line, apparently looking at each of the other gulls in turn before fitting in next to one near the end, which appears to accept the advance for it does not fly up. Then off goes the next candidate. I have watched this watery romantic version of musical chairs over the years and have never failed to enjoy it. Once I counted thirty-seven gulls taking part in the line-up, not near the islet this time but out in the widest part of the loch. Sometimes it takes place in the moonlight and I may be woken up by the cries at 2 a.m.

The process must also strengthen the social bond of the whole colony, for by this formal ritual all the gulls get a good concentrated look at each of their number, so helping them identify the other individuals.

Territoriality now increases – the slightest infringement of a nesting area by an immediate neighbour brings protesting calls and aerial swoops by the defending birds before they land back on their own little patch. One bird from each pair stays to guard their nest site while its mate flies off to feed. Then back it comes to relieve the other and take over as sentry. If there are prolonged late snowfalls, all the birds leave for the coast until the weather improves; then back they come to squabble over their nest sites all over again.

One sunny 21 April, as Moobli and I were touring the woods, and I had just heard the first call of the cuckoo (the earliest I have recorded it in the west Highlands), I realised the gulls' islet was strangely quiet. The birds were busy mating, although the sexes looked alike. The paired birds stood close to each other, the males making soft '*quer quer*' calls before flying up to land on the females' backs. With shivering wings or very slow flaps that were just enough to maintain balance, there they stood until excitement mounted then flapped their wings harder, pushing their rear sections down while the females pushed theirs up and the cloacae touched long enough to pass the seed. Often the males stayed where they were, calling and flapping gently, until a minute or so later the act was performed again. As I crouched behind the trees on shore I felt like a peeping tom! Certainly the females were made of stern stuff for they could stand the males on their backs for quite long periods.

Both birds help to make the nest, nothing much more than a scrape really, by turning round and round in the tufty tussocks or in tiny grassy patches between the rounded humps of grey and white lichened granite. Sometimes they twist grasses round with their beaks, and add others for lining.

Early one morning I was woken by a commotion from the islet. A pair of hooded crows had landed in the Scots pines above the nest sites, as if on a recce to see if any eggs had been laid yet, and the gulls were giving indignant abuse. Oddly, the gulls seemed to tolerate their presence later. This happened every April, as if the crows were innocently establishing that they perched in the pines all the year round and had a right to be there. They sat still for an hour at a time, maybe trying to lure the gulls into a false sense of security.

By mid-May the females are sitting tight, laying the two or three darkly lined and splotched eggs, whose background colours vary from khaki through olive green to brown so that the uninitiated could easily be led to believe three different types of gull had laid them in the one nest.

Over the years the gulls became used to our presence. Apart from sending dawn raiding parties to the bird table, they did not seem too upset when I boated over to keep the annual tally on the number of eggs laid and young finally reared. They just flew up in a cloud, calling occasionally, then landed to ride the water while I rapidly made my counts and left. The most eggs

– eighty-nine in all – were laid in the second year, after the wettest (yet fairly mild) winter out of seven, and thirteen young were raised that season. The lowest number – forty eggs – came in the sixth year, after the bitterest winter in fifteen years, and only two young were reared to flying stage.

Usually the egg tallies passed without incident but one June visit nearly ended in disaster – for me, that is. To add to my humiliation I had with me three young friends, up for a visit. We looked up at the end of counting to see our boat drifting forty yards away, heading down the nine miles of open loch in a stiff breeze. I had forgotten, when ramming the boat on to the grass sward, that four people getting out of it instead of one would make it far lighter. We were marooned! From the boat Moobli's broad bear-like head stared back with a comical expression which said: 'Did you *mean* it to do this?' Unfortunately I had never trained him to tow back a drifting boat. When one of my friends said she would swim for the boat, I tried to play the man of action, stripped to my pants and, heading for the least rocky part of my shore 200 yards away, plunged in. The shock of the icy water was so great I thought I would have a heart attack.

I ploughed on and looked up when I reckoned I was nearing shore to find I was only halfway across. I changed to breast stroke, amazed I was making so little progress. With horror I realised this was the deep-water spot where a fisherman had drowned. When I opened my eyes under water there was nothing below me but blackness. For a moment I did what one should never do – panicked. I gasped, swallowed water, began to thrash around with a dog paddle, sure I was about to drown too. Stop swimming, I heard in my head. I relaxed, concentrated only on surfacing naturally with minimum movements and taking breaths each time I came up, then set off again with slower strokes.

Eventually I felt the furry algae-covered rocks under my finger-tips and crawled ashore. Only later, as I changed in the cottage, put the small engine on the old sea dinghy and went to fetch the drifted boat, did I realise what had been wrong. Thoughtlessly aiming for the less rocky beach, I had been fighting the current from my own burn. Luckily the boat had curved into shore, where Moobli greeted me stoically, and I towed it back to rescue my friends. We had intended to explore a high mountain that day, and as I was still shivering with cold it seemed as good a way to get warm as any. We climbed to its near-3,000-foot peak, and found two rare flowers and an eagle eyrie which had been destroyed by a rockfall. On the way back the boat engine conked out and we had to row the four miles home. As we berthed the boat at Wildernesse, three gulls winged over from the islet and made chuckling noises over our heads before heading back. 'And you!' I muttered, barely able to raise my arm to make a profane gesture at their retreating forms.

Once the eggs were laid the gull colony posted its 'hit men' – sentries who kept watch from the outermost rocks and flew up to mob any predatory bird passing too close. They even came to call over our heads if Moobli and I walked along our own shore. By hiding in the east wood around dawn, then dashing across the intervening water in a good moment, the hoodies still managed to reach the islet unseen. There they hid deep in the tangles of pine needles, waiting for a change to swoop fast on an unguarded egg.

When Gilbert, the herring gull I had befriended on the sea island, flew over to renew acquaintance and help himself at the bird table, the colony greatly resented his presence in their area. The sentries saw him coming half a mile away and with loud 'keeya's tried to drive him off, but he dodged them easily. The common gulls didn't bother him at the table, but if he flew towards their islet upon leaving, they harried him away to westward. Gilbert made no attempt to purloin their eggs or young chicks and, once he'd had his fill, he leisurely beat his way back to the coast. I had the impression he often just came for the trip, because he was bored.

As soon as the young had hatched in late May and early June no large bird passing within a quarter mile of the islet was safe, even the large female buzzard which often flew over. Once as she flew high over the loch four gulls rose above the islet, calling as if discussing what to do, then two broke off, flew upwards and repeatedly dived down on the great mottled bird, which just kept soaring in circles, trying not to notice. The gulls followed her to the far side of the loch and over the hills until all five were lost to my sight. They also posted a far sentry, which sat on an exposed rock at about 400 feet on a ridge north-west of the cottage. If one of the buzzards or either of the crows flew above the west wood the sentry gull would give chase, making loud calls that brought two or more members from the 'Flying Squad' on the islet to lend it support. When I was indoors and heard a gull come screaming over I would know there was a big bird overhead – and I got one of my best photos of a flying buzzard in this way.

Occasionally the gulls would see off a heron. This huge, gawky, light-boned blue-grey bird which patiently fishes the shallows of rivers, lochs or tidal flats adopted the area around Wildernesse from the end of the first year onwards, coming in both winter and summer, despite having sometimes to run the gauntlet of the indignant gulls. Why it flew ten miles from the nearest heronry in a sea loch to fish our bays I had no idea. Possibly the weedbeds were prolific in prey. Once one of the sentry gulls dived over its head so repeatedly as it was standing knee deep in the next bay that it leaped jerkily into the air and flapped off to a more peaceful spot. Herons have been known to steal gull chicks now and then but I never saw this heron on the islet.

One 23 June, after a late season when the young gulls were just sprouting their first speckled brown feathers, I heard gulls calling out. Then I was sure I heard someone say '*Scram!*' in a harsh voice. I went out to find Moobli alertly staring at the shore; then I saw two of the gulls pursuing the heron beyond the lochside ash and alder trees. Each time they came near, the heron swerved to left or right with great aerobatic agility for so large a wing span, and cried '*Scram!*' in the same harsh tones. He seemed most indignant at being so chased, at having his dignified composure upset. It wasn't until the second year that an astonishing possibility occurred to me regarding this heron.

I was in the woods one June evening at dusk while a northwest gale was blowing noisily through the oaks and larches and saw the bird above the treetops, battling its way homewards to the heronry after its day's fishing. With its light body and large sail area it was being wafted all over the place, yet it persevered with deliberate slow wingbeats, gradually making headway over the ridges to the west. As it went my heart gave a little jump, for I thought I saw the outside toe of his left foot sticking out sideways. No, it couldn't be possible, I thought.

Five years earlier, when a pine tree falling in a gale had dislodged two young herons from their nest at the sea loch heronry, I had rescued one that was still alive and had managed to put it back into its nest. The heron, which I had named Harry, had just such a broken toe and had later adopted my own bay on the sea island for fishing. During the following years he had become reasonably tame, even fishing in the burn right by the croft. Could this bird, five years later, be the same? Could he have actually seen me boating up the fresh water loch, as perhaps had Gilbert the gull, and again adopted the area where I lived? It was surely impossible, I told myself. In the dusk I must have imagined that distorted toe.

I didn't have a chance to see the heron outlined against the sky again until 11 October, when he had taken to finding water creatures in the burn and maybe a mouse or a shrew in the lush forest debris along the banks. That day I again saw him fly overhead, this time in full daylight. I had my fieldglass with me and was sure I could see the toe sticking out. There was little doubt in my mind now that it was indeed Harry.

During the winters I saw him less often in the nearby bays, for as the burn and loch grew colder and his prey sought deeper, warmer waters so he spent more time among the better pickings on the tidal flats of the sea loch, where he then often had to join others of his kind. I counted seventeen herons lined up along that shore on one December day, all looking for water creatures coming down the burn that drained into the loch, or those which came up with the incoming sea tide. The final denouement of my experiences with Harry the heron is so extraordinary few would believe it had I not the photographic proof – but that story belongs later.

One early June afternoon, when the first gull chicks had hatched, I found the gulls faced enemies I had not thought of, for I saw an exceptional predator at work, and one against which the gulls had little defence. I was out in the garden, marvelling at the velvety black wings and brilliant crimson stripes of the first red admiral butterfly I had seen at Wildernesse, when I first spotted her. The falcon was flying along behind the ridges above the cottage, her scythe-like wings beating contrapuntally to the untamed rhythms of the summer hills and the grasses blowing before the north-west winds. Easy, just a stroke, a flicker here and there, mistress of the violent air.

Seeing a rock just over the ridge, out of the turbulence, she landed to survey the new scene that stretched below her with dark beautiful eyes, far keener than any man's: the little white cottage 1,500 feet below, the human in its garden, the curving woods of oak and hazel, the feathery larches fronting the shore of the loch, and beyond them the islet where the breeding gulls flickered like tiny specks of light.

Only eighteen inches long, she was yet the climax of evolutionary perfection, nature's complete killing machine. Her slaty-blue back reflected the bright azure of the sky, her creamy narrow-barred chest feathers were held tight to her body, and the black moustaches below her steely-blue hooked beak gave a solemn cast to her hunter's face. Now, from over half a mile away, she selected her victim and launched herself from the crest. A few strong beats and she was up again into the full vortex of the wind. She angled her wings back and began to move.

Only then, as I paused in my labours to watch the butterfly, did I see her arcing across the sky like a meteor, and at first thought she was a jet plane with swept-back wings. Down, down, down, she stooped across the blue curving firmament of the sky at incredible speed, then was lost to sight behind the east wood. I dashed down to get a clear view but was too late, for all I saw was her beating away to the east with a light brown feathery bundle in her talons – a fledged gull chick which could never have known what hit it.

I felt breathless, amazed. It was all over in a moment, a flash of time. The gulls, who mobbed and harassed any crow or buzzard that passed by, could never have seen her coming, or if they did, had no time to do anything about it. Now they flew distractedly, with indignant shrieks of fear and frustration, impotent against nature's lightning slayer.

To see the hunting flight of the peregrine falcon is to re-evaluate all of nature, for it is a primeval god.

Usually the peregrine kills birds on the wing, stooping down with such force it can snap off a pigeon's head with the strike, though it occasionally kills such fare as young rabbits on the ground. It was interesting that it could snaffle a semi-grown gull chick in this way, after locating its quarry from a distance. When I went to count them they always crept, as furtive as rats in daylight, under the tussock grasses or hid below rock ledges at surface level on the shore, keeping dead still, as if knowing a slight foot movement would send tell-tale circling ripples across the water. In the falcon's case, however, they would not have been hiding at all and didn't see her coming. Maybe the chicks had been preoccupied with the biting midges, now at their most numerous, for I often saw gulls and fledglings sitting on the rocks shaking the pests out of their plumage. The yearly counts of eggs and young reared to flying stage were interesting but I was disappointed to note a general decrease in the first six seasons:

Year	Eggs Laid	Young Reared
1	67	7
2	89	13
3	61	6
4	57	8
5	59	5
6	40	2

In the second and most prolific year, I also found seventeen cold unhatched eggs lying by the sides of nests, some crushed and stinking. Yet this was also the year the record number of thirteen young were reared. Causes of such failed eggs included prolonged cold spells, hail showers, careless younger parents and incidental predation. Some infertility may also have resulted from pollutive poisons in the bodies of the gulls' live prey.

Once, as we returned from a supply trip, Moobli stopped on his run and began nosing about among the rocks on shore. I boated over to where he had found a dead fulmar. This astonished me, for fulmars are truly birds of the open sea, gliding and wheeling for hours on end, resting on the waves and seldom going to land except in the breeding season. Yet here, on 28 June, was a dead fulmar on a fresh water loch nine miles from the sea. It measured eighteen and a half inches and its long slender wings spanned three feet eight inches. I sent it to Dr Douglas Ruthven at the wildlife analysis unit of the Department of Agriculture and Fisheries for Scotland at East Craigs, Edinburgh. Back came the report that the bird was in poor condition, its liver extremely pale and its stomach containing only a few beetle cases. Residues of organo-chlorine compounds were higher than in many seabirds but should not have been enough to cause death. It also had traces of oil at the bases of its body feathers, and I wondered if it had been trying to migrate from east to west coast via the inland fresh water lochs and had died from starvation after battling against the recent rain-filled north-westerly gales.

Over the years the gulls became more tolerant of my few visits and often came for food on the bird table. I felt they knew I was friendly, but I was surprised in late May in the fifth year when I went to count the eggs. As I walked carefully among the tussocks, with Moobli doing the same (his nose was useful for locating eggs that had rolled under herbage), and found the fifty-nine eggs from sixteen nests, I heard sharp whizzing noises above my head. Some of the forty gulls I had counted in the air were diving at me, and worse over Moobli, swooping down, just missing us and always coming from behind no matter which way we faced. Poor Moobli, scared by the sudden rushings of air by his ears, began to whirl to face the oncoming foes. We beat a hasty retreat.

It had been an extremely cold and snowy winter, and I thought instinct had made the gulls more protective than usual. Next day I glimpsed a stoat in my garden and later found stoat droppings on the islet, as well as several shells of eggs which had been carried up to the thick heather below the pines where the stoat had feasted on their contents. Stoats are good swimmers and clearly this one had visited the islet before – hence the gulls' increased aggression. The same thing happened the following year, when only two youngsters flew.

All this was natural predation. It was the first stoat I had seen at Wildernesse, probably a wanderer looking for a new territory, and as these animals seldom live longer than three or four years in the wild, and prefer rabbits or voles as food, I felt it posed little real threat. Gulls can live up to thirty years, and if all the eggs were lost for three years running it would not mean the end of the colony.

The gulls were opportunist feeders. If there were fishermen or boats along the river banks where the main weedbeds were, they flew to the sea shore. One year, when they all left the islet on a spring morning before laying, I couldn't find them on the shore or on the weedbeds. Coming back, I saw a new field had just been ploughed up by a farmer – which disappointed me for the area had been a nesting ground for curlews, oyster catchers and other birds – and was now covered by common gulls eating the new rich pickings. They had soon found the finest new food source in the area, as had other gulls, lapwings and a few curlews too.

Further south in winter the migrant gulls depend on ploughed fields for much of their food. Conservationists need to be cautious before accusing farmers of often being enemies of wildlife, for by making ground more fertile they help many species. Here some of the gulls were even courting, running round each other with opened wings behind the farmer's moving tractor. They would thrive on the rich invertebrate food thus unearthed and get into good shape for laying fertile eggs. Weeks later I saw three times the normal amount of birds including curlews and oyster catchers, still feeding on the field. The farmer had only ploughed a few acres in this instance and the shift to slightly higher nesting grounds was minimal.

What perplexed me in late June and July was that while the young were well grown and thus needed more food, the adult gulls spent even more time on the islet than they had earlier in the season. They were also moulting some of their wing feathers. Now and again I saw small parties of gulls flying closely round some of the bigger lochside trees, weaving and diving as if trying to land on the branches, but unable to do so because of their webbed feet. What were they wasting valuable energy like that for?

I found out one 29 June when I boated to the islet for the year's chick count. As I picked up one chick with half-inch-long tail feathers it suddenly regurgitated the contents of its stomach into my hand. All caterpillars – which I later identified as those of mottled umber and winter moths. Next time I saw the gulls wheeling round trees I stalked nearer and put the fieldglass on them; they were actually buffeting the branches with both their feet and wings. This dislodged the caterpillars, which promptly let go in fright and dangled below the foliage on their long safety lines – and the gulls were swooping to gulp them up to feed themselves and their chicks. Both mottled umber and winter moth caterpillars can accrue in plague proportions some years and will completely defoliate trees like sallow, birch and oak, so that some of the trees die. The gulls had not only found a close food supply but were performing a useful forestry service too.

Often as I trekked about on my main golden eagle studies, I saw small parties of gulls floating airily over the high hills, sometimes above 1,800 feet. Here, too, they were seeking out the caterpillars of various moths which fed on the many grasses and heather. They had found yet another easy food source, one which only occurred in summer but for which, again, they did not have to fly as far as the sea coast.

In early July their interest in the food which I put out also increased, but only to feed themselves. They never took bread or scraps back to the islet for their chicks. Shortly after dawn they began to fly in, land on my iron roof with scratchy clanks and loud shrieks of '*kisha*!', which began diminuendo and increased to fortissimo. Then there was a silence as they looked about to make doubly sure the coast was clear. As I lay abed, there were brief *whrrr whrrr whrrrs*, a flash of white angelic wings as they hovered about the table, seizing a piece of food before *beat beat beat*, and away they went, their light cigar-shaped bodies jerking up and down slightly between their long wings.

It was not until the sixth summer that I found they were exploiting a new food source. Coming back by boat, I saw one flying oddly, jerking this way and that, actually hovering at times and striking out here and there with its beak, its wings shining in the sunlight like those of a tern. It was catching flies. It seemed that they were quick to learn from each other, for eight days later I saw three gulls doing it together.

On 18 July in the fifth year I found that even though the chicks were well grown and taking their first flying exercises, they were still not immune from predators. I was walking through the shore-side alders when I saw the big female buzzard launch herself from the trees in the east wood. She took a deep uncharacteristic fast dive down, flying like a great brown hawk, just missed one of the chicks, realised her attempt had failed, then got the hell out of there as the 'Flying Squad' went after her. She was moulting, I noticed, and had several flight feathers missing. Maybe her mate had been a bit lax in bringing food for their youngster that day and she had gone out to look for a quick meal. As the gulls chased her away she let out an odd squawking '*kraow*' noise, occasionally turning on her side and indignantly thrusting out her talons to try to grab the angry swooping gulls. She sailed over the highest ridges and the gulls broke off singly leaving one gull, maybe the 'on duty' sentry, to keep diving at her until she was out of sight. I had little doubt she had accounted for a chick or two, probably in poorer light.

By mid-July the grown chicks are left more and more alone, spending their days in flying exercises, scudding round the islet in a changing cloud of grey-brown specks, swooping down upon the water and up again, practising water landings and take-offs. Well before dusk one or both parents would come back to feed their offspring. Occasionally a gull would bring its youngster over and perch it on the chimney, as if telling it to watch as she tried to catch crusts I threw into the air. A few more days of longer flights, now more often with adults, and by late July (the earliest and latest dates I observed over the years were 18 July and 2 August) all the gulls have left the islet for the seashore to the west.

By the sea the young watch their parents and learn how to survive and find food in a marine environment. In the harshest months they seem to migrate inland and further south, joining black-headed gulls to feed on marsh lands, river meadows, farm fields, and even recreation grounds and tips on the edges of towns. But there is one phenomenon I find fascinating. Every September, usually around the 23rd, and then again in late November or early December, a flock of common gulls, with several still browny speckled youngsters among them, come beating east to west along the loch shore, often circling back over the cottage too and flying low over the islet. Then on they go, loosely following the line of the shore, but flying easily with slow beats, shifting hither and thither a few yards each side, as if relaxing into the wind rather than fighting it in strong direct flight.

I am sure they are my little flock coming to check their islet, and feel cheered to know that when the blankets of mist and curtains of rain lift to reveal the first blue skies of February, they will be back once more, long before the cuckoos, to usher in another spring.

Chapter 8

The Loch in Spring

When one lives alone on the shores of a large wild loch spring is the magical time of year, for the water mirrors the increasing blue between the passing grey clouds, reflects the green of new leaves, the shimmering wings of breeding birds arriving from spending winter further south, and it magnifies the music of their courting calls.

For the last seventeen years I have lived in wild remote places, always close to the sea or long fresh water lakes, and life in a country cottage without this unique blend of water, land and sky would not for me be the same. I often think of the multitude of benefits that come from living by water: you can drink it, cool down with it, and wash, swim, dive, boat and fish in it. One winter no doubt I would be able to skate *on* it. Above all, water adds a multi-changing beauty to the landscape.

I have long learned that to appreciate such beauty at its best one must move quietly, dressed not in the bright safety colours of the uninitiated but in natural hues that blend with the water and landscape so that one can mingle more closely with the wildlife; that it is best whenever possible to remove the petrol engine and set out on the loch's broad breast by almost silent oar power. Sometimes, for the sheer peace and beauty of it, I have rowed the full eighteen-mile return journey to the village at the loch's western end.

The first of these journeys took place in early April, when after two cloudless days the third also dawned in a sky of clear amethyst, the early rays of the sun sending a slight misty vapour from the glassy calm surface. There were no other boats on the loch. It was too early for the tourists or summer fishermen, and the keeled dinghy glided along in a silent watery paradise. On the flat marshy meadows seven lapwings had arrived from wintering further south, and already two were making courtship flights – flying high on throbbing wings then diving madly earthwards, twisting as if out of control and calling '*peeoowit wit wit, peeoowit*,' then pulling out again just before crashing into the ground.

There were three hooded crows on the mudflats, respectably attired in smart grey waistcoats, and one was pulling out a worm after listening for it with head cocked to one side like a thrush. One dived with twisting flight among the peewits and two turned to chase it, following every manoeuvre of the crow until it shot up high and turned off. It seemed more like a game of tag than malicious pursuit at this time of year for once the hoody had gone, they chased each other. When the lapwing eggs are laid from mid-April onwards the males keep guard over their incubating females and do their utmost to harry away the marauding crows.

Curlews, later nesters, were picking their way through the grasses, undisturbed by the silent boat, but skulking behind tufts the moment they sensed eyes upon them. Then on they went again, to probe deeper than any other wader with their long downcurved bills. Occasionally I heard the high trilling bubbling song of the males as they glided over their future small nesting territories, to which they were already staking claims. Enchanted by the balmy silence that enshrined the wild lonely calls, I shipped the oars and let the boat glide.

To my right, three whooper swans swam from behind Heron Island, a wild place where the whitened trunks of ancient pines lie like bleached prehistoric bones among the new tangles of willow and small alder. It seemed they were all that remained of last year's family for the two white adults, carrying their heads more erect than mute swans, their wedgeshaped bills a far paler yellow, steamed along in front while the bird in the rear was greyer, obviously the one remaining cygnet from the three to five eggs its mother would have produced. As they passed I heard them making soft short noises, '*owoo*' and '*whooaw,*' like muted dog barks. Soon these three would be on the whoopers' extraordinary 500-mile non-stop flight to Iceland or north Scandinavia where they have to lay, incubate and rear young relatively fast for such big birds in the short fine seasons so far north. Their incubation period can be less than thirty-five days – up to nine days faster than the golden eagle.

Whoopers ceased to breed in Britain in the eighteenth century, but in recent years a rare pair or two have nested in the Highlands, though I have never known of any nests on my loch. They cannot breed successfully until four or five years old, and as they live up to twenty years or more they also spend some of their last years without breeding. So most of the few whoopers still seen in the Highlands in summer are non-breeders which have not made the long northerly flight, being either too young or too old. I had no wish to startle the majestic white birds into flight and waited until they were further away before reaching again for the oars.

As the long blades flashed and dipped making little sunlit whirlpools, and the boat glided round a small gravelly peninsula, I saw a party of five goldeneye ducks with their comic conical-shaped heads paddling above a half-grown weedbed in an oval lagoon. They saw the boat and *plop, plop, plop* dived alternately, evidently feeling I was far enough away to make flight unnecessary. I was halfway across the lagoon's entrance before one by one they emerged again, after a good half minute underwater. Expert divers, they feed on small water creatures, so when goldeneye stay on a stretch of water it is a sign there is plenty of life down below.

As I saw them resurface I smiled for it was from these quaint little ducks I learned the most amusing wildlife stalking method of all. Late one March I was approaching a hill lochan at 1,200 feet when from a distance I saw a single goldeneye diving from its surface, staying under a few seconds, then coming up with bits of weed in its beak. There was no tree or bush cover, just a few rocks; how could I get close enough for photos? Then I realised it couldn't see me when it was under water, so the second it dived I ran like hell and threw myself behind the first rock.

The goldeneye came up, swam about, then dived again. Off I went like a middle-aged arthritic hare to the next rock. Once more this happened and luckily I took two 'insurance' pictures, for in the middle of my dash to the last rock on the shoreline the duck surfaced, saw me gallumphing forward, and took off!

It was after chancy stalks like this – which can only be effective if you are fast enough to duck and freeze the instant the water shows signs of the bird resurfacing – that I sometimes saw goldeneyes performing courtship rituals. The dumpy black-and-white drake swam round the browner female, flashing his circular white cheek patches, flicking his head back against his body with his beak sticking straight up into the air, and then lifting head and neck high as if trying to be a swan, and swelling out his jaw feathers so that he looked as if he had mumps. A pity, I often think, for such effort may be wasted as goldeneyes fly to northern Europe and Scandinavia to nest, where no doubt the drakes go through it all again.

It is likely, however, that many mate while still in Britain, so that by the time they get to their breeding lakes the females are almost ready to lay and finding their usual tree-hole site is first priority. For goldeneye to nest in Britain is rare, though a pair nested in Cheshire in the early 1930s. When, in 1960, Royal Society for the Protection of Birds' workers noticed the ducks were staying longer in late spring round Speyside lochs, they started putting out special nest boxes, of a type Finnish and northern Swedish farmers used to collect some of the birds' down. At first only jackdaws, starlings, tawny owls, squirrels and even wasps used them, but under Highland Officer Roy Dennis the RSPB persisted. In 1970 a duck raised four young from a box in Inverness-shire. Eight years later a dozen females were using the boxes, and the following year the birds laid in twenty-one sites, fourteen females were successful and 110 young were hatched, including two broods from unknown wild sites. This is all very encouraging, and any landowners with suitable sites wanting to help goldeneye in recolonising Britain should contact the RSPB.

By then I was passing below a long wood of oak, birch, hazel and holly trees, and suddenly a pair of mallards rose almost vertically from where they had been dabbling for water-plants, unseen by me. They seemed to flap harder than most ducks, wings well back, making little headway for such effort. Once under way, they travelled fast, long necks extended and looking like mini-prototypes for Concorde. As I looked up I saw two great black-backed gulls cruising high above the shore. Usually sea cliff birds, I had noticed a pair of these huge 27-inch gulls, our largest, came ten or more miles inland in winter and early spring, looking for dead sheep and deer whose carrion provides them with the kind of feast they rarely find on the seashore at this time of year.

Their light yellow eyes are possibly as sharp as a buzzard's, and once they find a carcass they monopolise it until gorged. Being larger and more aggressive, they are capable of keeping even a buzzard waiting in nearby trees until they have flown back to their shore roosts. Now this pair passed overhead with slow but powerful beats of their great black white-edged wings spanning up to five and a half feet. One whirled round in a tight spiral and began to glide down. Instantly the one behind noticed, flew to join its mate, and they vanished behind a small ridge on shore. I'll look there on the way back, I thought.

I hauled rhythmically onwards, not dipping the oars too deeply, letting the natural leaning weight of the back do most of the work, with a switch of the wrists at the end of each stroke and leaving a small curl of water behind the stern. Suddenly I saw something struggling in the water to my right. I braked by sinking and pushing against the oars, and went to look at a fine Emperor moth, one of Britain's most beautiful and common in my area. Newly hatched, inexperienced at flying, it had tried to cross the loch and had not made it. Its thick antennae proclaimed it to be a male. I put it on the seat where it flopped wetly, turned right way up, quivered its wings dry in the sun for three minutes, then sprang into the air and flew off. This time it made the shore safely.

As I turned into the last north-west leg of the loch, six miles done and three to go, my nostrils were assailed by an acrid scent – smoke. I turned to look – the whole hill above the village, from the sea to far ridges to the south, was afire. Holes of bright vermilion, lower banks of smouldering red, and bursts of orange were blazing in pockets all over the mountain sides for over a mile, while above there puthered globes of smoke, forming a dark umbrella pall above the entire village and surrounding hills. The slight breeze was blowing from the east, from the village to the fire, so at least property and life seemed secure.

When I hauled into the shallow bay below the village and walked up to the post office, a small group was watching the fire. Pockets of it were glowing redly only sixty yards beyond the village, the grasses and heather twigs seeming to scream and whine as they were being consumed by the flames; the noise and smoke reflecting the pain of the dying wilds. Fine black dust from burning peat that had taken thousands of years to form was settling over everything.

At this time of year many crofters and sheep farmers indulge in 'muirburn', where they try to burn off old long heather to make areas for new grass to grow for grazing, but no one seemed to know how this fire started.

'Two of the shepherds began it, and it got out of control,' volunteered one man.

'Could have been an accident, by the sun through a loose piece of glass,' said another whom I knew to be a crofter.

'Glass, my eye!' replied the first. 'I'll bet they were out there with matches. And when it got too big they buggered off!'

'I see the vandals are at it again!' said a bearded man, coming to join us from where his car was being repaired at a garage. I felt surprised to find someone living locally who felt strongly against the burning of the hill since most of the folk I knew seemed to go along with the tradition, and I agreed with him it was a huge risk to take, just in the hope of securing a little increased grass growth. 'Och, there's nae point burning that hill off,' he added aptly. 'There's not a bloody sheep up there anyway.'

A Land Rover with a fire pump and coils of hose pulled up and three jerseyed men leaped out – the fire brigade from a nearby village. Faced with the raging hill in the near distance and no water supply, there was nothing anyone could do but ensure the fire didn't creep back against the wind through dry below-surface roots to the village area.

Such out-of-control fires are not uncommon in the Highlands in April, and every year hundreds of square miles go up in smoke. Young regenerating trees, wild flowers, orchids, ground nesting sites for birds, and large numbers of small creatures – lizards, mice, voles, frogs, and even rabbits and weasels who may escape the flames yet are asphyxiated by smoke in their burrows – are destroyed. If the fire travels fast enough birds too perish, and with late fires (I've known them in early May) their eggs or young are also consumed.

The fires certainly do burn off heather but unfortunately nearly all of it, young growth too; yet they hardly affect the underground rhizomes of the pervasive bracken weed, which ultimately takes over large areas and is poisonous to sheep and cattle. On the steep high hills of the northwest Highlands burning often destroys so much vegetation that wind and rain erosion of soil is accelerated and only bare rock is left. Indeed such burning only makes sense if done in smallish areas at a time in strict rotation. Heather has a life span of fifteen to forty years and at most should be burned off no more than once every ten years. Dr Jim Lockie, formerly head of the Department of Forestry and Natural Resources at Edinburgh University, and other research scientists too, have recommended that anyone with ten acres might burn just one acre each year, and preferably not an acre adjacent to that burned the year before. With a hundred acres, ten acres a year can safely be burned but in separate lots of one acre each, again not immediately next to those most recently burned.

In this way, ideal from the whole ecological view, it is possible to maintain a balance of new grass and nutritious young heather, medium heather which provides shelter and food for many species and which is *still* good for sheep, and some old longer heather, giving bad weather shelter for sheep and deer as well as for grouse which need it for moulting cover. Grouse populations in the west could perhaps be partly restored to their former high numbers if such burning rotation were widely followed.

Of course controlled burning demands much human labour and that costs money; it also requires intelligent planning. Perhaps for the hill sheep industry, which can only survive by subsidies anyway, this seems to be asking a lot. Yet it is obvious that haphazard uncontrolled burning, where scores of square miles can be destroyed by a single foolishly applied match in the spring when winds are at their most unpredictable, is little short of selfish, short-sighted, irresponsible stupidity. Nor can the extensive loss of soil on steep hills benefit any farmer.

These fires often cause damage to Forestry Commission wood-lands too. Over recent years letters have been sent from district offices to landowners, occupiers and tenants in their areas point-ing out that muir burners are required under the Hill Farming Act 1946 to give notice in writing to their neighbours and can be held responsible for damage to neighbouring lands, and offering in some circumstances to bulldoze fire-break strips between the sheep grazings and the woodlands. Some hill farmers, however, are quick to resent any interference from outside, and a week after the dispatch of letters in my area tradition was asserted and several big heather fires were burning, a few causing damage to Forestry Commission plantations.

One farm manager in the region legally notified the Forestry Commission that he and his workers were going to burn a small area: the fire got out of control and went into the forestry block. The fire brigade from the nearest big town, fifty miles away, was called in, a helicopter dropped pumps, every able-bodied man was rounded up to help. Even so 2,000 acres were destroyed. The golden eagle pair in that area, which had three eyries, did not breed at all that year. A few years previously another fire started by the same estate workers actually burnt out one of the eagles' sea-cliff nests.

More recently there have been indications that the best sheep men realise the ecological problems and do their best, though it is far from easy to control the burning properly.

On the afternoon of the local fire I finished shopping and back-packed the supplies and two gallons of paraffin down to the boat. Before the nine-mile row home again I thought I would wander down to the bridge over the river that flows out of the loch. On the way I saw a small party of whooper swans feeding on the weedbeds where the loch narrowed into the river. To reach the more succulent shoots some up-ended vertically, splashing with their big dark webbed feet as they tried to push their light bodies further down.

As I looked over the bridge at the changing water patterns below, the river crystal clear and flowing smoothly as there had been no rain for two weeks, I felt I had little hope of seeing any salmon. Although they began to come in from the sea here in small numbers from March onwards, I believed the first runners – which can include the largest fish – never showed themselves, travelling at night through waters seldom more than three feet deep. Suddenly I saw a sharp movement, as if a piece of weed had detached itself from behind a rock. It was a big fish, and it darted with a swift tail swirl behind another rock, where it sheltered as if having a rest in the back-curling waters in its lee, its greeny-blue back merging perfectly with the strands of weed near it. It was slimmer than a fresh-run salmon ought to be, I thought

– probably an old kelt, a fish that had survived spawning in one of the burns that fed the loch and had made it back to the sea. Now it was coming back for a second time. I must have moved slightly for there was a flick of gravel and it was gone, up, under the bridge.

As I rowed home again, falling back rather than hauling on the oars, a slight north-easterly warning me to take it easy as I would probably have to fight hard up the last third of the loch, I almost forgot to check where I had seen the great black-backed gulls. Then I saw one perched on a shore-side rock. As it took off, its mate rose in the same instant from behind the small ridge and flew off behind it. I beached the boat and went over. With their powerful beaks, they had made a hole in the neck of a dead hog, a year-old castrated male lamb which appeared to have succumbed in the last snowfalls, and had pecked away much of the face flesh. I knew it was not a fox's work as there were no throat wounds, none of the usual scats, and Moobli could obtain no fox scent. Over the years I had learned that these great gulls are voracious carrion eaters and are often quicker than buzzards or crows to spy a dead lamb. They kill and eat other birds, including waders and smaller gulls, and have been known to half devour shot-crippled birds before wildfowlers can gather them.

Reckoned by some folk to be a cruel bird for this reason – though no animal can be cruel in the sense that man, with his intelligence and foresight, can be – these big gulls have sometimes surprised me. In summer a pair occasionally came down to snaffle a piece of stale bread I had thrown out, but as soon as one picked it up and flew shorewards, the other immediately followed and was allowed by the first to share it, both birds pecking it in turn. Few other gull species would share food in this way, especially out of the breeding season.

To lighten the load for my arms and back, I dropped Moobli off on shore, to run the three miles back to the cottage. Because he always minded his own business, these weekly runs of his disturbed wildlife hardly at all. Occasional sheep, which he had been trained to ignore, would start to panic as they saw him coming but he took pains to make wide detours round them. After the first year hinds and calves learned he was harmless and would stand on a low ridge, long ears pointing forward nervously, watching in case he *did* decide to go for them, but not so nervous as to cease chewing the cud.

As he drifted along, tan legs flashing like quicksilver, pairs of red-breasted mergansers paddled out temporarily from rocky arbours where they were already prospecting nest sites in any hollow with rank concealing herbage. I learned over the years there were three pairs of these colourful diving ducks nesting between Wildernesse and the bay beyond Sandy Point. Their numbers seldom varied, despite the arrival in the third year of a new river bailiff who decided there were too many mergansers in the area for the fishing interests and began shooting them in May.

Red-breasted mergansers and goosanders have long been protected in England and Wales where they are uncommon but, believing them to take a heavy toll of salmon par and young trout, Scotland's lobbying fishermen succeeded in having them excluded from the Protection of Birds Acts 1954 and 1967. They could be shot, even in the breeding season. In fact, it has been proved that both birds eat fewer salmonid-type fishes than other prey, and I was glad to see they were removed from the list of 'pest' birds, which could be killed by landowners and occupiers only, in the 1981 Wildlife and Countryside Act. Such new protection under this Act, however, is still largely cosmetic for 'authorised persons' can still kill any wild bird, except those in Schedule 1, if they show the action is necessary for 'preventing serious damage to livestock, foodstuffs for livestock, crops, vegetables, fruit, growing timber, or fisheries'. The accent is on *serious* damage but there is still too much leeway, in my view, as there is in the entire Act.

Home again after the eighteen-mile row, I staggered out, legs stiff, and childishly proud that while my hands were sore in places there were no blisters. But, oh! my poor backside. The skin was sore from rocking to and fro on the seat, despite the use of a cushion picked up from the truck on the homeward journey. Hot, I took a deep breath and plunged into the water, still icy from winter, yelled with shock, then dashed up the hill for a towel. What a grand day, I thought, as I rubbed myself dry. Moobli, however, seemed to feel the fun was just starting. Although he had run-walked three miles, maybe swum two more, he bounded up and down, exhorting me to hurl hefty sticks for him to chase.

On the next supply trip ten days later I saw the tops of the hills were black where they had been burned, and a high wind was fluming away the ash and dry soil, making small dust storms on some of the ridges. Later, when I hiked to the start of the fire area, I was surprised to see new grass was growing where patches of heather had been – just what the burners had intended. I found out, however, that while it looked nice and green against the black patches, the new growth was only common molinia grass, containing very little nutritional value. A week later the first bracken shoots were coming through, invading slowly into areas where they would have no competition for several years from heather, nor from the few sparsely scattered young trees that had survived the cropping teeth of sheep and deer, for those now stood as stark little black skeletons of burnt death.

By mid-April the white-fronted geese which have been wintering among the mainly residential greylags on the marshy wetlands at the end of the loch become more restless. Just before dawn they fly to riverside fields and other small marshes at burn mouths up and down the loch. They seem to be looking for the finest grasses and plant shoots to build up the final strength for their long flight to breeding grounds in Greenland, up in the Arctic tundra. White-fronts also nest along Russia's Arctic coast to Siberia.

One 11 April, after two days of constant light drizzle on a warmish south-westerly wind had melted snow blizzards, I wandered sleepily down to the shore, where I was soon startled fully awake by a loud whooshing whistling of large wings. Just beyond the shore fringe of ashes and alders it looked as if the contents of a farmyard had taken to the air. A small flock of white-fronts, their black chest bars standing out, their white foreheads easily seen, rose almost vertically from the areas of thick new sprouting grasses. As they levelled out and winged away down the loch, a few making gabbling cackles, I cursed for not having taken more care – and a camera. There was a neat long line of greenish droppings, thicker than pencils, over the belts of green, as if they had been careful not to foul the whole area.

Four days later I heard a musical chuckling sound, like a party of music students singing snatchy squabbly phrases from some comedy operetta. I went out to see a flock of sixty white-fronts heading way over the hills to the north-west, as if on course for Greenland. I thought that was the last I would see of them that year but, two days later, I was sneaking through the marshes when I saw an even larger flock grazing in the wet meadows. They seemed split into small family groups and occasionally one goose would lower its head and make metallic noises at others, as if it felt they were grazing too close.

Some farmers reckon geese to be a nuisance, competing for grass and clover with sheep and cattle. But providing the flocks are not too large the early grazing can help the plants, for they then put out double shoots – which makes for lusher grazing a little later on.

That year I saw a large flock flying from east to west down the loch in a shallow V formation, at about 2.30 p.m. on 25 April, the height of each bird altering slightly. I rough-counted 132 birds before they vanished towards the horizon. At 5.45 p.m. another flock of ninety-four went down the same way, and I was sure these were birds from lochs and grazings further inland, which would have a final 'fill-up' on our marshes, the last before the sea.

That was the latest I had known the white-fronts to linger, but it had been an extremely wet winter in the Highlands. Maybe they had a way of knowing that, if the winter had been bad here, it would be up in the Arctic too, for these geese need to be nippy about their breeding. They seek thawed-out ground where possible, for snowfalls can occur at any time, and it takes a month for the eggs to hatch, another week before the young can run after their parents freely, and many more weeks before they can actually fly. Snow also brings down predatory Arctic foxes which steal eggs and young if they get the chance, yet the male white-front gander is a good watch-dog and with help from another bird can usually drive off such predators. It has been found that almost 90 per cent of eggs hatched are reared to flying stage, though why these geese don't breed in the Highlands is, for me, a mystery. Probably suitable areas are just not undisturbed enough.

Among the ducks, the mallards seem to be the earliest nesters. One 30 April, a warm but drizzly day when the loch lay heavy and grey like polished slate, Moobli, who was then not quite six months old, found our first mallard nest. We were walking along the shore of Heron Island when he paused, one heavy outsized paw raised like a pointer, and sniffed towards a blunt knoll covered with bilberry leaves. Keeping him back, thinking he had maybe found a carcass, I stole through the undergrowth.

Suddenly a dark brown mallard duck took off, winging away with flashing purple and white wing patches. She was already sitting on eleven light green-grey eggs, and the nest, which was actually on top of the knoll, giving her an all-round view, had as much brown down in it as an eider duck's. She had not had time to arrange the down over her eggs as she would have done when leaving them normally. As she skimmed low over the water in a wide circle, watching us, I left hurriedly. As we boated away, she made another circle, this time right over our heads. Then she slooshed into the water by the island. When she was satisfied we were far enough away, she flopped up the rounded rocks and pushed through the vegetation to get back on to her eggs. I didn't bother her again.

One little bonus that day was the first Greenland wheatears arriving after their long migration from Africa, crouching in dry rock crevices not far from the shore like little black-masked

highwaymen, as if regaining strength to fly again a few days later
to where they breed in Greenland. It always amazes me such tiny
creatures can travel so far when humans worry about driving a
tenth of such distances.

By now the red-breasted mergansers were courting. The handsome male with his red eyes and beak, glossy bottle green head with
its double crests, white neckband, red breast and white-flashed
wings, swam near the drabber female. She often swims lower in the
water, making him look larger and giving him a patriarchal air as
he bears down on her, bowing and raising his head while giving a
soft '*cauw*' note. Then he may raise his crests, lift his wings and
drive round her so fast he sends up spray like a tiny speedboat.

Once she starts incubating her eggs (she can lay up to thirteen)
she will sit as tight as a woodcock and won't move unless you are
almost treading on her. She can see you, peering through the
thick herbage which screens her nest, but you have a devil of a job
to spot her – or even the nest if she is not on it. I have only found
two merganser nests in ten years, though I don't specifically look
for them.

A late layer, any time from May to early July, the female
merganser is one of the best mothers in the duck family, for it is
she alone who makes the nest, lining it with grass, leaves and her
own grey down, and she who does the incubating. Then the decorative males have time on their hands. On 2 May I saw two swimming along together, a week later a third had joined them, and on
5 June I saw five male mergansers together on the gravelly spit of
Sandy Point. They were like proud fathers gathered for a bachelor party, while their wives did all the work back home. They
were certainly enjoying the warm sun – and who wouldn't after
long underwater swims in such cold water. Mergansers are
carnivorous, and catch fish in their saw-like bills by diving down
and chasing them. They can live and nest on fresh or sea lochs
(where their favourite food is sand eels) and there seems little
reason why they should not colonise more of the south, except
that they like large undisturbed areas.

By mid-June we see the first merganser families in the water – mother followed by up to a dozen anxiously bustling light brown ducklings which stick tight and follow her everywhere. If the boat startles them unexpectedly round a bend, the duck pretends she is wounded to lure the danger away, while the young wriggle away like snakes in a hurry, their breasts churning up the water, their little feet paddles flailing madly. When they are older, the mother flies off in a circle, and the chicks dive one after the other as if in a water chorus line. It is amazing how far the little mites can swim, always away from whichever direction the boat moves. But they do suffer high mortality. Black-backs, herring gulls and even sparrowhawks and the occasional otter can snatch them from the water's surface. Foxes, stoats, wildcats and owls can take them on shore at night, and if the duck has reared three to flying stage in six weeks, by mid-August usually, when they are left on their own, she has done well.

The last of the small summer breeders to arrive – I have never seen one before 26 April, and once, after the coldest winter in sixteen years, they did not arrive until 3 June – are the common sandpipers. They are worth waiting for. These small wading sprites, under eight inches long and not even as big as starlings, embody the summer spirit of the loch shore. I go down day after day, wondering if perhaps this year they have fallen victim to those 'sportsmen' who kill small birds round the Mediterranean. Then one day after a long dry spell, the loch as low as it has ever been, the first rain comes, the earth smells sweet after sucking it up and in an hour one hardly knows it has rained at all, and as I step through the gate arch of logs I hear a shrill mournful piping '*wee peep, wee peep,*' and a little brown long-tailed bird with white underparts is flickering from rock to rock, pausing now and then to bob its head and flick its tail up and down like a wagtail.

What a curiously unique flight has the sandpiper – a few shallow rapid wingbeats, *brrt brrt, brrt brrt,* so fast they appear to be twanging, then a short glide with downcurved wings almost touching the water as it skims in a half circle, then it lands on another rock or in shallow water and probes for insects or larvae, or picks flies off stones. They start courting within a few days of arrival. Though both sexes look alike, I sometimes hear the brief shrilling song of the male, then see him chasing his mate in twisting turning flight, both showing their white bellies. Then she alights and he skims round her, lands near, shivers his wings and bobs his tail, showing the outer feathers richly barred in chestnut and white. This display usually ends in a ground chase, their heads held low and with him haring after her, their slim green legs going so fast they become invisible so that they seem to flow over the little rocks and stones.

Once I saw a courtship flight which ended in tragedy, oddly enough on a Friday 13 May. I had just beached the boat when my two sandpipers flew past at great speed. The female in front was dodging and twisting low between the rocks pursued by her amorous mate, and as I slung my pack on my shoulder I thought I heard a sudden thumping thwack. I looked round, saw only Moobli behind me and one of the birds flying round in a circle. Then it landed with shrill '*swee de tit*' cries, as if objecting to our sudden arrival. I paid little more heed, thinking maybe Moobli had stumbled over something, and went on up to the cottage.

Next morning I found a dead sandpiper lying behind a rock some twenty yards from the boat. It had a broken neck. Clearly the thump I had heard was the sound of it hitting the rock, over which it must then have fallen out of my sight. Such an accident must be a million to one chance. The bereft male hung about the shore all summer, piping, as if also regretting the accidental results of his ardour. He didn't find another mate.

The following May he was back on the shore with another mate – unless they were two entirely new birds – and this time I found their nest, purely because one of the birds sprang out of the turf and ran over the ground trailing a wing as if it was broken, a ploy used to lure enemies away. The nest, a small natural bowl lined with grasses, contained three small pear-shaped beige eggs splotched with reddish brown streaks. As I was busy with eagles that year, I did not set up a hide, but in late June the birds were obviously feeding young for they ranged far and wide for the large quota of insects they needed and finally came round the cottage. Maybe they found insects attracted by the scraps round the bird table, for they tweedle-dummed through the grasses, thin legs going like windmills, darting their long sharp beaks everywhere. Once I found a sandpiper in the porch. I had never known them so tame.

On 3 July we were by the boat runway on the shore when Moobli found a scent, followed it into a tangle of thick grasses and bracken, made a whuffling snort through his nose and stayed still, nose deep in the herbage, his tail wagging. He had scented out a sandpiper chick. I had never photographed one but after catching it I found it ran off so fast a reasonable photo was not possible. As neither parent seemed to be around, I took it to the house and photographed it through a hole in a box, then released it where Moobli had found it. It ran like a tiny ostrich straight into the bracken near the boat. Five days later I saw it running along the beach with one of its parents. Presumably it was the only survivor of the three eggs.

One summer phenomenon of the loch has been the increasing number of cormorants coming inland to fish. For the first three years I saw scarcely any, but during the fourth July I looked out of the window to see a huge black bird outlined against the sky, bigger than any goose, which circled the garden on long broad wings, then plunged into the loch. As it swam about, twisting its hook-beaked head characteristically from side to side, I was surprised to see it was a cormorant. Flying so close, its size had been impressive and my books told me the bird's average length is three feet and its weight 6½ lbs. I am sure I have since seen bigger cormorants, huge males flying low over the waves (because they lessen the speed of the wind) and showing the silky white patches which grow on their thighs during the breeding season.

This cormorant adopted the rocks of my boat bay as a diving perch for two days but must have decided it was too close to humanity for safety, and it took off for the gulls' islet where it was apparently tolerated by the colony for the gulls made no moves to drive it away. It seemed they knew it was not interested in filching their eggs. Since then I have seen as many as three standing together on the islet, wings held wide open to dry in the sun-filled breeze, and quite ignored by the gulls.

As I have watched these incredibly fascinating birds over the years I have felt transported back into prehistoric days for they are descended from the long-extinct pterosaurs, flying reptiles with twenty-foot wing spans, and the toothed diving birds of the Cretaceous era 130 million years ago. They emerge from the depths, green eyes glinting, their short thick feathers glittering green and violet like the scales of ancient fish, and as they stand motionless, clinging close with reptilian feet, the mist swirling about them, they look like gryphons from another age, carved from the very rock on which they stand.

Cormorants have to dry their wings because they are the only diving birds which don't produce waterproofing oil on their feathers. At first this seems a curiously inefficient quirk of evolution, until you realise the wetter feathers increase their weight and so help them to swim more efficiently under water. They can alter the specific gravity of their bodies at will, by letting out air from their lungs and air sacs, for when alarmed they sink lower so that all you see if the surface is choppy is the neck and head sticking up like a scrawny periscope, held upwards at forty degrees, before diving to escape a boat. Or else they will take suddenly to the air, rising easily from the water for so large a bird. Strong fliers, although not strictly migratory, Scottish birds have been found to disperse as far as 330 miles from where they were bred. Some ringed English birds have been found in Portugal.

Cormorants catch fish by outswimming them, with powerful drives of their strongly muscled dark legs and webbed feet. So efficient are they that from the sixth century Japanese fishermen trained them to hunt fish. Harnessed to lines, with a neck ring to stop them swallowing their prey, an expert can fish with twelve cormorants at a time. Fish weighing up to 2½ lbs and eels as long as themselves have been found in cormorants' stomachs.

Cormorants will prey, but not often, on young waterfowl and will catch and eat rats and moorhens. Once an eleven-inch-long kitten was found in one bird's gut. Their throats can stretch enough to swallow all such prey, their windpipes being protected by strong cartilage rings like those of a flexible vacuum tube.

For many years they were on the 'pest' list of birds which could be killed by landowners or occupiers, largely because they were believed to predate heavily on trout and young salmon. In 1965 Dr D.H. Mills published a study of the cormorant in Scottish inland waters, mostly those in the Conon basin and Ness river system, in which he found that on lochs alone the food from twenty-one cormorants consisted of 42.8 per cent brown trout, 19 per cent young salmon, 38.1 per cent perch, 9.5 per cent eels. But out of all the areas – lochs, rivers, estuary and firths – the figures for eighty-five birds were: 24.7 per cent brown trout, 21.1 per cent flounder, 16.4 per cent salmon, 12.9 per cent perch, 10.5 per cent eels, and 2.3 per cent pike. While trout and young salmon figures seem high, the destruction of eels, perch and pike removed much serious predation on the salmonid fishes. Dr Mills pointed out that while the cormorant may affect the brown trout population, it is not a serious predator of young salmon. By removing some brown trout it may actually benefit the salmon, as brown trout themselves predate upon young salmon. Control of cormorants, therefore, may only be worth bothering about if cormorant numbers on trout lochs are high.

Be that as it may, the cormorant was removed from the 'pest' list in the new Wildlife and Countryside Act 1981, so it now enjoys ordinary protection. Not that this means a lot, as with mergansers and goosanders – owners and occupiers can still kill any wild bird, except those in Schedule 1, if they can show the action was necessary to prevent *serious* damage to crops, growing timber or fisheries.

Since the fourth year I have regularly seen cormorants on the loch in summer but never more than half a dozen at a time. These birds are non-breeders for the cormorant breeds in colonies on sea cliffs and islands, and only very few have been found to nest inland. On summer nights the great birds roost on the huge standing dead spars of ancient pines on Heron Island, and it is quite a sight to see them floating and wheeling down from the sky towards dusk to land, despite their webbed feet, on the dead white branches.

As they stand outlined against the darkening firmament, black wings held open like cloaks, they look like a group of draculas, and it does not need too fanciful an imagination to believe that, come the darkness, they may all be off to fulfil some sinister visitation.

Chapter 9

The Loch in Summer

With our usual arrogant way of looking at nature purely from the view of our own benefit, we seldom admit – though we should because few estates undertake restocking programmes – that the most serious predator of adult trout and salmon is man himself. Nor can I exempt myself completely.

I had always loved fishing in my youth and had fished for both sport and survival on the Pacific coast of Canada and on the Scottish sea island. When I first came to live at Wildernesse, being one of the few who fished the loch to have his rights legally inscribed with the Register of Sasines, Edinburgh, I naturally felt I would spend many summer days with rod and line – and, at first, I did.

In late May with the loch warming up, the first may and caddis flies taking to the air, small brown trout infiltrated up the burn to live in the deeper pools below the waterfall. They did this to become the first fish to exploit the new insect hatchings and the new food being washed downstream from the miles of glens above. And usually in early June, when the loch surface was pitted with small circles from fish rising to take the new flies, the returning sea trout and salmon making splashing leaps after insects, I would boat out to the best fishing rocks near long sandy bays and try my luck above the sparse weedbeds.

Despite my years of sea fishing and coarse river angling in England, dapping with dry fly was for me a new art, though I thought I could soon master it. I was wrong. To dap you need some forty feet of silk floss on your reel, with about five feet of nylon trace tied to it and at the end of that the dry fly. I made most of my own flies from small feathers, trying to imitate the colours of whatever insects were then hatching. I would then putter, or preferably row, in a stiff breeze until a hundred yards off shore and let out the line. The trick is to let the breeze take out as much floss as possible, the fine silk strands of which catch the wind, then with dipping movements in slow semi-circles the fly can be made to dance over the water in a realistic way, maybe thirty feet or more from the boat.

At first it was I who did most of the dancing, mentally with rage, for the fly often fell too far into the water, got wet, turned into a little black unappetising lump, and had to be dried. Or a slackening of wind sent the floss into the water, soaking it and making it too heavy to float out when the breeze returned. It had to be wound back, dried and sprayed with silicone before I could try again. The thin strands stuck on everything possible, the rod's tip, tiny projections on my old rod's bindings, a nick in the boat, even on to calloused skin on my fingers. I could not dap at all on calm days, when wind was gusty and unreliable, in rain, in bright sunlight when fish could see me easily, or in east winds when the air was colder than the water and the sea trout refused to rise.

For two years I caught no fish by dapping at all and was shocked that I missed with every strike. In the third season I learned the difference between a genuine bite and just a 'swirl rise' – when the fish suddenly shoots up, hits the fly so you think he's taken it, and you strike into empty air. For in this type of rise the trout only hits the fly with its body or tail. You're raising them to roll against the bait, but you don't catch them. Even the cormorants, drying their wings on the islets in the July sun, had their beaks open as if laughing at my efforts – though of course they do this to keep cool.

One day at this time I rowed all the way to the truck, trolling a salmon spinner behind the boat. Near a reed bed the rod banged, I grabbed it, felt a surging dive and run, began to play what was surely a salmon – when the line went limp. Hell! I reached the truck with my arms aching and had to admit the rowing muscles were losing stamina. As I drove up the farm track to where the road ended I was hailed by an old couple I had seen fishing the loch.

'Aye, many's the time I made my bracken bed down by your place,' said the man, an old age pensioner. 'So I could be back at the fishing rocks early in the morning.'

I was about to invite them to visit, after so ponderous a hint, when he showed me their creel, stuffed with over a dozen sea trout, for their deep freezer! I didn't invite them. I don't like greed, I told myself. Even *innocent* greed. Actually, I was probably jealous too. Two-week summer visitors or not, they knew more about fresh water loch fishing than I did.

Eventually I got the hang of it, learned to strike with only a light wrist flick and, as ever, let the fish tire itself before landing it. My limit, however, was two fish in any one session. The sea trout I caught never weighed more than two pounds but that was enough to last me two weeks, and all I could keep fresh anyway.

One early July there were an exceptional number of parr, the young salmon beautifully striped with butterfly blues before they turn into silvery smolts in their third or fourth year and go on out to the sea. Everywhere I rowed I saw them flashing away from the boat. Often they came close to shore or into my boat bay, and the burn pools and small bays were filled with rippling circles as they sucked down tiny insects. I wondered if this had been caused by the exceptionally dry winter before last, when there had been far fewer spates to smash them up in the torrents or scour away the eggs in the gravel of the river beds.

When the 1976 summer drought hit Europe it even affected the notorious Highland rainfalls too, and by 24 August my burn was a mere trickle. The loch held many whitish flecks of growing vegetation and algae and even began to look like stagnant water. No boats came up; all the fishermen were complaining of no fish and of the river being so low the salmon were not moving up it at all.

It was at that time the new ghillie of the estate on one side of the river decided he would dig out some of the weedbeds in the hope of making new spawning places for salmon. As the big digger machine and its driver rumbled out over the dry river bed to start work, the ghillie and head of the fishing syndicate which owned the rights on the other bank turned up and asked what was going on.

'How long have you been here?' asked the ghillie from the far bank. 'Since April,' said the new man.

'Well, I've been fishing here for forty years,' was the reply. 'And *you* think you know this river?'

The new ghillie, somewhat abashed, not wanting to get on the wrong side of the established locals, said he would abide by their opinions and invited the syndicate head to give his own instructions to the digger driver.

'Right,' he said. 'Get that digger out of there and take it away.' And that was that.

A few weeks later the new ghillie told me he had dug a trench on his side of the river a mile further down, heaving the gravel to one side to make a deep pool and an artificial bank. The idea was to create a new pool which would hold salmon in low water for the fishermen to try for, and the new gravel bank could be used by the fish later in the year for scooping out their redds.

Of course the depletion of Atlantic salmon stocks, of which Scottish rivers have the lion's share in Britain, has many complicated causes. At the time of writing, the Association of Scottish District Salmon Fishery Boards revealed that total catches had fallen by 25 per cent in the preceding five years but, again, the decline has been going on for many years. Long gone are the days when three rod and line fishermen on Grimersta's lower loch landed fifty-four salmon weighing 314 lbs on *one* August day in 1888, and ended up with 333 fish for the week! And of course, too, the remedies lie far beyond more Scottish estates indulging in restocking programmes. International cooperation is needed, especially regarding the great modern toll of the fish taken by the high seas fisheries of Greenland, Denmark and the Faroese in the main sea feeding grounds of the North Atlantic stocks.

In the 1800s many main rivers in western Europe and North America were developed for navigation, high locks for ships to pass and weirs were built, and spawning gravel shallows were dredged out to help navigation. In this century came pollution by toxic wastes from the new industries along the river banks. Hydro-electric dams and water storage schemes altered natural flow and drowned out many spawning shallows. Rivers flowing through forested country are far richer in food organisms so the vast forestry fellings along many streams, together with the deoxygenising waste discharged by pulp mills, also had a damaging effect.

In the 1950s the West Greenland fisheries were only catching enough salmon for their own relatively small human population. But between 1960 and 1964 catches swelled from about 100 to 1,500 tonnes. Ships from outside Greenland then began drift net fishing further off the Greenland coast, so that by the early 1970s the harvest had grown to 2,000 tonnes a year from the area. It was found through tagging returns that this was the main sea feeding ground for salmon from all western Europe and the Baltic and North American Atlantic coasts. The International Council of North West Atlantic Fisheries was formed and gained some control of the Greenland fisheries. Later its role was largely taken over by the EEC's fisheries policy, and in January 1982 a convention in Reykjavik laid the foundations of the North Atlantic Salmon Conservation Organisation, with headquarters in Edinburgh, through which the EEC, Norway, Sweden, Iceland, Canada and the USA would try to cooperate on rational management – including conservation, restoration and enhancement of the stocks. At the time of writing, catch quotas for the West Greenland and Faroese fisheries were to be dealt with under separate agreements, however, and positive results were not yet known.

The complaint of many estate and fishery owners, along with government scientists, in the past has been: what is the point of full blown salmon 'ranching' (breeding and freeing smolts into the seas in the hope they will return to the same tanks or rivers to breed when mature) if most of the fish was bound to be caught by the high seas fisheries off the Greenland and Norwegian coasts?

It is my own belief that it is the responsibility of all salmon nations to enhance the wild stocks in any way possible. One key to this is surely through the proliferating salmon farming industry which, when all research is completed, might perhaps be required to breed a certain percentage of smolts (disease free, of course) for release into the wild as a condition of operating at all, and to do so under licensed supervision. The initial biological problems of salmon farming have been largely overcome; by mid-1982 there were sixty-two different farms, mainly in the Highlands and Islands. In 1980, 600 tonnes were produced, 1,200 the following year and some 2,000 tonnes in 1982.

Of course, this more than made up for the decline in wild fish for sale as food and brought the price of salmon down considerably, to the further dismay of the legal netters who in Scotland have charge of a £5,000,000-a-year industry employing 1,200 people. While it may seem all this should reduce the incentive to poachers, commercial poaching continued at a steady level. As the rod and line men always insist, both vigilance and penalties should be stepped up. But rod and line men also have a responsibility towards wild salmon stocks and until and unless the situation improves, they might come off their usual high horse position of saying: only a few are taken in the rivers by rod and line – most go through. The sport fisherman who catches one, two, then tries for a third in one day is as guilty of greed as the poacher he rants about and upon whom he seeks to put most of the blame.

Be all this as it may, the new ghillie came up to me in the following December, his face wreathed in smiles. He said he had that day seen salmon digging their redds in his new bank and felt vindicated. Certainly, running a salmon river is no easy business.

I only fished the open loch, however, and when not dapping with dry fly, I was trolling for salmon. Here a swivel, a small weight, and an artificial lure are tied to the line, which is let out some fifty yards until the lure is a fair distance below the surface, and then towed slowly behind the boat, travelling in shallow curves to keep the lure out of the slight turbulence from the prop. Ideally, one should row. One can also tie a couple of flies a yard or so apart and tow them along the surface in the same way.

In seven years, probably because I didn't fish much, I caught only three salmon, all by trolling, the advantage of which is that you can just sit in the boat and relax. Like many fishermen, I suspect, I was not out there just to catch a fish but to be out in the wild, to know moments of peace and beauty and, in my own case, to glimpse yet another aspect of the wilderness in which I was lucky enough to be living all the time.

Over the years, however, I came to realise I could have such idyllic moments without fishing at all. Besides, as I sank myself deeper into wildlife, became increasingly occupied with exciting studies on species like deer, fox, badger, wildcat and eagle – apart from the time I had to spend trying to earn a living pittance as a writer – sitting in a boat with rod and line began to seem boring by comparison. This slowly induced a somewhat philistine attitude to the traditional summer pursuit, and my earlier love of fishing waned. In the end I seldom went out to fish more than half a dozen times in each summer.

Perhaps catching my third salmon had something to do with it too.

It was a gold and gorgeous late July morning, with a cool northwesterly breeze reducing the sapping heat of the sun, but as I puttered round the blue lagoon by the fishing rocks, the wind fell. The day too calm and bright now for dry fly, I unsuccessfully trolled a silver spinner, had no luck either with pootling along various reed banks with a bluish spinner designed for mackerel, and went back to the centre of the main pool – where my little old fishing engine immediately cut out.

Here, facing a deep half moon bay of shallow rising sand, shielded from the main loch by a rocky promontory on my left and two islets, the wind picked up briefly, but only in small gusts that ruffled the surface and not steady enough to use the dapping floss. I looked in my lure box. There hung a large bright white and yellow wooden plug I had made for pike in boyhood. It had bright orange cheeks, black eyes painted on yellow irides, and a metal scoop under its chin that made it wiggle through the water.

I had made it for muddy Sussex river waters where its glaring colours would show up and induce the ferocious pike to strike. No fish in its right mind would take it in so clear a Scottish loch, I thought. Well, I would see if it still worked anyway, and tied it on with a 1½-oz weight four feet above it. Such plugs are meant for casting out and spinning back on the reel and their beauty is that they float up and so keep clear of most weeds, even though the weight may be amongst them before it is reeled back. Well, I would see how far I could cast it before restarting the engine and trolling it across the lagoon on my way home.

Holding the line free on the multiple reel, still sitting, I slung weight and plug out about forty yards, then let the weight carry it down while I pulled the engine starter rope. For three pulls it wouldn't start, but on the fourth it sparked and we were under way. The lure could not have travelled more than a few yards when the line went tight. Damn, I've snagged a rock or clump of weed, I thought, and cut the engine again.

Suddenly there was a heavy boring downwards, as if the plug had hooked a rock off an underwater shelf and the rock was now heading towards the loch bed. I gave a slight flick of the rod hoping to dislodge the hooks, cursing the fact I had probably now lost my ancient plug. Then the pressure eased again but as I reeled in there still appeared to be something heavy on the line. As it came nearer the boat there were occasional twists that caused the rod to dip a foot or more, strong but slow, ponderous wrenches. I thought maybe I had hooked a clump of weed *and* a fish, so that its struggles were being muted by the weight of the weed. Such a fish would see the boat and would struggle again, and so I let it go but still kept pressure on the line. It only went about five yards but again the pull was slow and heavy, then I began hauling it back once more and it travelled slightly to the left, solid, unyielding, as if I was towing a submerged log.

Then a pale light appeared in the water and my heart leaped as I saw what it was. A huge salmon, about three inches across the muzzle. One hook of the plug had caught into the lower edge of an ugly downturned mouth, the great black eyes stared up at me from the depths and I saw the underjaw was curved upwards in the kind of ugly kype male salmon get in the spawning season. Down it went again, with a loop of its body and twist of its broad tail, slowly, as if it was tired. Before it vanished my eyes took in the long blackish body, a thin body really for the size of the fish.

It bore away for some twenty yards, making slow lunging heaves and I let the top of the rod take them. Soon I was reeling it in again and once more the great head appeared. I saw the fish had a faint white streak between its eyes, the place where possibly fungus would grow and finally consume him after spawning. I lifted the rod to turn the head and hold it just above the surface and reached instinctively for my gaff hook. As I saw the hopeless beaten look in those huge dark eyes, the spouts of water forced up into the air by the old gills, I was suddenly overcome by a strange sadness. Here was an old mended kelt, a fish that had returned to spawn in the burns at least one season, had made it back to the sea and now had returned again.

What right, I thought, have I to wrench this poor old fellow, whose flesh would taste poor anyway, who was too large for me to keep fresh for eating, from his natural world? He put up a courageous fight for a longer life, beyond that of most of his kind, for a second chance. Was I not a mended kelt too? I thought suddenly. Had I not, after a childhood among nature, followed by years as a journalist in many of the world's big cities, been given a second chance to return to the way of life I had always most valued?

I let the rod tip down and watched him lying under the surface, gulping to recover more strength. I could see myself in this old fish and I knew I would let him go. He was only hooked in the lower edge of the operculum. I slid the gaff down the line after turning it round, inserted the tip into the edge of the hook and, with a gentle downward poke, flicked it out.

'Go on, you old bugger,' I said, waving my hand over his head.

Down he went, with a series of sideways lunges as if he thought he was still hooked, then realised he was free. As I watched the last downward drive of his tail, I thought what a master of the world this salmon has proved himself.

Hatching from an egg laid by his mother in gravel up one of these burns into an alevin or 'fry', he had wriggled up through the redd and had survived by feeding on tiny insect larvae and nymphs when 95 per cent of his relative alevins would have perished. Within a year he had become a parr, with lovely blue blotches on his sides, and had drifted down the burn into the loch where he would have spent two to three more years. Then, turning silver, he had become a smolt and had migrated down the outlet river to the sea. There he would have spent from two to four more years mastering the salty ocean deep, ranging along the coasts of northern Europe to the main feeding grounds off Greenland.

There he had dodged the sea nets of the Greenlanders, Danes and Faroese, then back he had come, scenting for the pheromones of the very river or burn in which he had been born. He had survived too the legal (and illegal) netters of the Scottish sea lochs, had found his own native river, had successfully run the gauntlet of the rod and line fishermen up it and had traced his home burn, maybe even the exact redd in which he had been born, often after boring his way through water less than a foot deep. There he had fought off other males for the right to spawn, to release his fertilising milt over a hen fish's eggs.

Then, when nearly all salmon die after spawning, this mighty fish had fought his way back out to sea and had mastered the whole process for a second time – maybe even a third. Few humans, even today, could be as well travelled as that old fish. Maybe I was being sentimental but as I boated home, ready to face the tyrant typewriter again, I did not regret releasing him. If ever any creature had earned the right to pass on his genes, he had.

Late summer on the loch passes in many days of watery haze, the '*quawrk*'s of the mergansers carrying over a mile across the unruffled surface, the youngsters learning their first hunting dives among the weedbeds near the shore. On still days the air is filled with midges, every cubic inch having its own vicious tenant, and slow outdoor occupations like gardening, winter firewood cutting, repairing boats or painting the house need to be restricted to windier days when the little pests take refuge in the herbage.

From the flat waterside marshes come the mournful '*coorli*' of curlews, and as you pass close in the boat the birds, whose youngsters are not quite to flying stage, run and hide behind tufts of grass or clumps of rushes, to stay by the chicks. By early September the first small flocks of pochards wing in from northern Europe, and I find them floating buoyantly like dumpy corks, the drakes' handsome chestnut heads conspicuous, below the ash trees on the shore. For a day or two they spend most of their time sleeping, heads back under wings, a few sentries staying alert to give any alarm, as if needing rest after the long flights.

Through winter the small groups join up and by late January I have seen flocks up to thirty-two strong, but always wary, swimming away from the boat at a quarter mile – perhaps because they really have to batter their way over the surface with wings and feet a fair way to become airborne.

While the pochards are one of the first reminders of coming winter, the mid-September gales are the next. I woke one night to rain lashing on the tin roof, with the occasional tinkle of hail-stones too; the sudden gales making the porch roof rattle and bounce despite the logs I had put on it to hold it down. I went out in the morning to find the loch had risen a couple of feet. Pounding waves had filled the big boat, which I had neglected to put on the trolley and haul further up, and shoved it against an alder where they were still breaking over it. The oars and wooden runways were fifty yards away, also banging against rocks and trees. Luckily the alder's roots had prevented the boats being beaten on to rocks. Another bailing battle lay ahead.

Every year I have the feeling these first real storms are out to end my existence. One mid-September day in the third year I battled up the loch in the small boat with most of my gear (I had been away on a work trip) against strong easterly gales. I had beaten them before and thought I could this time. I was wrong. The engine cut out twice, wave after wave came into the boat and down it went, with me in it, luckily not far from the shore.

Once, when I was trying to winch the half-full boat out as the waves refilled the stern, the entire heavy winch assembly snapped off the wooden post under the strain. It whizzed past my head so close it clipped the edge of my left ear. Two more inches to the right and the local coroner would have had an odd case on his hands.

These are the gales that begin to strip the leaves from lochside trees, but at least they clear the midges from the air too. By 23 September or so the temperature drops noticeably and within days – though I try to discipline myself to last out until 1 October – I have to light the first fire.

It was on the last day of September in the seventh autumn at Wildernesse that the extraordinary end came to the experiences with Harry, the heron whose life I had saved as a young nestling over nine years before. As I walked down to the shore in the morning with Moobli, he put his nose down as if receiving a strange scent. Then he stood, ears cocked, peering forward, tell-ing me as he had been trained to do that there was something unusual ahead. Moving carefully, I went down the steps below the log archway, and suddenly a heron shot up from right beside the old sea boat and flew, but slowly, a mere fifty yards to a small bay to the east, where it landed again with an awkward folding of its long blue-grey wings.

I had no chance to see the tell-tale broken outside toe of the left foot which would tell me it was Harry. Hoping to find out, and get a good picture, I dodged back behind the lochside ashes and alders and tried to stalk him with the camera from above. But I over-estimated the distance and, when my head appeared over the bank above him, he spotted me and took off round the shore edge of the east wood where he was lost to sight through the leafy trees. I stalked more carefully across the burn, peered through the bracken and located him amid the shore-side tussocks. Hissing Moobli back, I crawled nearer and got a photo of him partly obscured by grasses. Then he was off again, gliding to a bay thirty yards on, and I thought I could see the toe sticking out. Now there was nothing between us but open land and further stalking was impossible. Having much work to do, I went back to my desk.

Next day, the incident forgotten, I went out early and saw the heron again tucked into the grasses right by the boat. Our sudden appearance made it try to take off but it tripped on the rocks and fell over, its wings held wide apart as if in supplication. It was clearly very weak. I hastened over and caught it gently by the lower neck, taking care to keep my face and eyes away from that dagger beak. Just as Harry had done as a young nestling when a falling pine had dislodged him from his nest and I had found him trapped by a leg in a bush, so this heron tried to peck. It was thin, breast-bone sharp, its gullet and crop empty and it was extremely frail. Tucking its body carefully under one arm, I examined its feet.

It *was* Harry, for sure; the outside toe of his left foot was hard, unbending, and twisted out at right angles.

Here was confirmation that the heron I had occasionally seen in the area since our second year (I had hardly believed my one sighting of the broken toe) had indeed been Harry all the time. Now, in his extremity, he had clearly felt a real instinct to seek help from me. Why else would he be there two days running, below the only human dwelling in fifteen miles of roadless loch, and right by my boat? And, after being scared away by our sudden appearance the first day, why else would he have come back to the very same spot on the second?

Another striking fact was that he had not sought the new boat, now up on the trolley, but to lie beside the old boat, the one I had used at the sea island and had rowed round the bays he himself had adopted after flying the nest, the very boat he had swooped low over on the day I had left the island. He had no doubt often seen me rowing the same boat up this fresh water loch too. He obviously knew the boat well, and that it was mine. Once could be accidental. Twice could not.

Yet if by some strange deep instinct he had sought my help, he was still terrified when I actually came close and he pecked at me, albeit weakly. I felt his beak; not razor sharp, as when young, but blunt. His claws too were old, grey and blunt and his plumage seemed pale with age, lacking the bluey sheen of the heron in its prime. By this time he would be nine years old. Perhaps he was dying of old age.

I carried him up to the cottage. Having no fish and knowing herons will eat mice, voles, even rats, I fed him some small lumps of meat, talking soothingly. He stopped pecking then and made no more attempts to stop me touching, stroking and holding his long neck while easing the meat down his long elastic gullet with gulps of water. He just squatted on his rear knee joints, his long legs weak, blinking his eyes slowly as if finding he had to accept what was now happening to him, but knowing he was in good hands.

A sick animal needs peace at such a time, so Moobli and I went for a trek, covering ten miles of loch shore and back over the high hills, keeping an eye open for a frog or two to give Harry. For the first time, when coming through steep tussocks, I actually fell down twice. It's not only Harry who's getting old, I thought. I stalked two groups of stags and hinds but in the poor light took no photos. When we got back Harry was looking even worse, his head and neck stretched across the floor. He had brought up all the meat given him. We hadn't found a single frog so I tried to feed him a few even smaller pieces of meat but again, with weak movements, he brought these up too. It seemed he was too far gone; his stomach could not cope with food any more.

I decided that if he was dying I wouldn't let him go in a human kitchen but outside, with the open sky above and the wild scenery of his life about him. I fed him some warm bread and milk in the hope he might at least be able to digest that. Then, so that a prowling fox would not attack him, I put him among the long grasses of the now unused wildcat pens a few yards from the cottage, with a deep bowl of water and some meat beside it in case he recovered.

When I went out next morning he was dead, his eyes half closed as if he had not wanted to leave without seeing the sky that had borne him on all his travels between the lochs. As I gathered up the limp and wasted body and folded the 5-foot 3-inch span of his wings, I felt sad, remembering the times we had shared in so strange a relationship, especially on the sea island. He had lived a full life, had certainly reared some young, and from the moment I had found him dangling from his left foot in the bush below the heronry and had overcome my fear of heights to put him back into his nest, he had survived well over nine years. I read somewhere a heron could live as long as twenty-four years but I was sure Harry had died of old age – life can be tough for such big birds on Scottish lochs. That he had returned at the end, apparently to seek help from the human who had saved him once before, was an enormous if tragic compliment, and my only regret was that he had left it too late. It was a poignant experience.

Every year at this time autumn advances slowly, but by mid-October the chill in the air has come to stay, the hazy mists of summer are seen no more. When the sun does shine now, it burns with a fierce, pure, untrammelled heat. Too soon such brief 'Indian' summers are over, purple rainclouds gather from the south-west and the loch turns a steely ice-dark blue. By the end of the month the whooper swans are flying back for the winter on singing wings, followed in days by the white-fronted geese, sometimes in skeins fifty strong, all honking and changing positions as the horizon recedes behind them – all to winter on the mossy marshy wetlands at the end of the loch.

From now on, I force myself yet again to accept stoically the isolation of the cold winter ahead. As I plod away at the old typewriter to the moanings of the wind, the crashing of the loch, I am consoled by knowing that as far as my brief human life is concerned, there will always be another spring.

Chapter 10

The Magnificent Divers

It is in early spring that the real 'stars' of the loch arrive from wintering on the sea coasts further south, the great black-throated divers.

These odd birds – twice as rare as golden eagles, three times as rare as peregrine falcons, for there are only about 140 breeding pairs in the whole of Britain – are wary, hard to study but immensely fascinating. For me they embody as no other creature does the wild spirit of the loch.

Heftily built, some twenty-eight inches long, when flying high on their short but powerful wings they look like arrows of twanging steel. They can dive to great depths, travel a quarter of a mile underwater, outswim their fishy prey, sink low when paddling along by emptying their air sacs, adapt to both salt and fresh water, and are extremely solicitous when taking care of their young. Their magnificent summer plumage, with its rich blend of slate greys, blue greys, purples, blacks, creamy underparts and sooty throat patches must be the sleekest of all water birds', and their array of intricate white neck stripes and snowy wing bars, if you're lucky enough to get close, dazzle the human eye. No wonder these rarities enjoy special protection. It was a full five years before I dared apply for a government licence and work them from a hide.

On 9 April in the first year, the hottest of nine cloudless days, I was perplexed by some mournful high-pitched mewing wails from across the loch. Then came harsh '*powoo, powlyow, powoo, powlyow*' gobbling calls, louder than from any ducks I knew. I thought these sounds were made by mergansers. I went out, raking with my fieldglass the far loch shore three-quarters of a mile away. There I saw two stout grey necks held upright above rounded blue-black humped bodies that floated almost as low in the water as cormorants'.

I soon identified them as black-throated divers from my books but didn't see them again until 2 September when a pair drifted by as I looked through the window by my desk. They moved fast but effortlessly, without any churning movements, and appeared to make a wider detour when they saw my boat on its runway. How shy they were, I thought.

On 12 May the next year, the day before a BBC television team was due to arrive to spend a week making a film about my work, I realised I had not yet done the yearly count of the common gulls' eggs on the pine tree islet. As I boated over I saw a diver leave the small eastern lagoon, steaming along like a small barge, yet with barely a ripple, before it dived and disappeared.

I had just concluded the count when I found an odd depression sparsely lined with dry white grasses among the scattered stones, a mere four yards from the nearest gull nest. There was a distinct flattened path from it to the water's edge. Realising this could be where the divers were going to lay, I swiftly left the islet. Of all seasons to choose! I thought. But the television crew cooperated, made wide detours round the islet in their boats and agreed not to film on it or even on the mainland opposite it.

Occasionally, from a small tree-lined bay a quarter of a mile away, I checked with my fieldglass and twice saw one of the divers sitting on the nest. In early June I was checking again when I saw a diver swimming west of the islet, and there by its side, bobbing along so that at times it was obscured by wavelets, was one brown downy chick. I resisted the temptation to go on the islet, for the family might still be going ashore there for rests and sunning themselves, and I didn't want to do anything to discourage them from breeding there another year.

My hopes rose on 14 April the following season when I saw one of the divers swimming past the islet. Two days later, boating for supplies, I saw the black-throated diver near the sandy point three miles away. Its head kept turning as it looked up into the sky. I followed its gaze: there, high above, another diver was flying north-west across the loch, its creamy belly flashing in the sunlight. There was no snaky down-bend in the neck that makes it so easy to identify the red-throated diver in flight.

To my disappointment it flew right over the swimming bird and I watched it disappearing through a cleft of the high mountains to the north-west. As occasionally I searched my area of the loch over the next six weeks and saw no sign of divers, I was sure it had either failed to recognise the terrain and had missed seeing its mate, or was a diver from a loch further north anyway. Certainly a sad error may have occurred.

On 3 June, however, I saw a pair swimming near the green island five miles west of my home. It couldn't have been my presence that made the islet pair move for there were far more boats and people at that end of the loch. Maybe they didn't like the continual racket kicked up by the gull colony. On 10 August I saw not only the green island pair again, but flying overhead towards the east a second pair which slooshed into the water halfway towards the sandy point – probably the pair which had nested on my islet the previous year.

From then on I established there were two pairs summering on the western ten miles of the loch. On 24 September I saw one diver with one chick, which had been able to fly for a month and was almost as big as its parent, near the gulls' islet. But they had not nested there. That was also the latest I ever saw black-throated divers on the loch; usually they had flown to sea by early September.

In the fourth year the nearest pair arrived back on 24 March, the earliest date ever, oddly coinciding exactly with the return of the common gulls. Above the gulls' raucous clamour I could also hear their high weird mewing wails, and that day I was also

treated to the rare sight of their courting display. It was a windless afternoon, the sun hazy so that the loch surface had a gimlet-
grey milky colour.

Suddenly one of the birds lowered its head to the water, extended
its short pointed wings upwards, made two slow motion flaps, then
arched them beautifully so that the wing tips touched the surface.
This was followed by a brief slooshing chase in which the female in
front made shallow splashing dives, as if turning over in the water.
Then both birds swam round each other, threw their heads on to
their backs and turned sideways so that their creamy underparts
shone across the water like beacons. One of them lifted one thick
webbed foot out, then paddled hard with the other, driving itself
round in circles, while the other preened its own back feathers
through its big dagger-like bill. After that I kept well away from the
islet, where the gulls were now screaming over their little territories.

Tantalisingly, two days later I heard the divers calling loudly,
not with their usual mewing wail but a far more metallic and
strident '*keh-kerlyee, keh-kerlyee*,' and again I felt there was no
more unearthly bird sound one could hear on Scottish lochs than
this. The calls, only two, came from near the islet, but as soon as
I went outside with my tape recorder they stopped. The birds
must also have dived for I saw neither of them. One diver
appeared near the islet on 19 May but again they didn't nest
there, nor did I see any young at all that year.

In a sort of desperate hope I applied the following March for
a Nature Conservancy Council licence to photograph divers on
the islet, feeling such optimism in itself was almost a guarantee
they would not nest there again! On 21 April, as I was watching
from my own shore the common gulls mating, I heard some
raucous '*rawl*' calls from far down the still loch. Was it the divers?
I had not heard their sounds for a year and memory was unsure,
but when I swung the fieldglass all I could see was a seedy-look-
ing pair of mergansers on the misty surface.

As I was driving along the farm track on 1 May after anchor-
ing the boat, I saw the year's first diver – floating fifty yards off

shore, head tucked back under its wing as if exhausted from a long flight. It was a full six miles west of Wildernesse. My pair probably won't even come back this year, I thought pessimistically. But as I was busy working with eagles from a hide on a high cliff thirty miles away, and I had little spare time, it seemed maybe just as well.

A few days later, however, I saw a black-throated diver swim slowly from the islet and paddle steadily down the loch. I waited until it was out of sight, then rowed over. To my surprised delight there was a new nest scrape in almost the same place as three years earlier. This time there were four common gulls' nests within five yards, each of which contained three eggs. Three driftwood twigs had been pressed into the ground beside the divers' nest spot, as if the birds had been too lazy to remove them. Excited, I made a quick survey of the high rocky area beneath the stunted pines where I could put up a hide, noted the main vegetation was good length heather, and left the islet swiftly.

My main hide was up on the eagles' cliff. I made a new one from bent hazel wands and green plastic garden netting with a tight sacking front, and after checking through the fieldglass neither diver was there, I rowed over quietly with it. I was halfway across when one of the gulls' outlying sentries flew over with loud calls, giving me away. I had forgotten them, thinking I could sneak unnoticed on to the islet from the nestless side in the dying sunlight, screened by the pines. Yet the gulls knew me of old and when they saw I was not walking among their nests they soon settled down.

I set the hide up between the pines and among the long heather on the edge of a rocky ridge, which gave a superb view of the divers' nest. I dared not go down and check again. The dry heather was lush and thick and I soon worked it into the hide netting so I would be completely hidden. Then I set the camera tripod inside and left. Ideally I should have had a helper or two to put me in and out of the hide, so deceiving the birds into believing that when they had left no one else remained behind.

My only helper lived thirty miles away and was already working overtime from pure kindness putting me into the eagle hide. I was sure the physical characteristics of the islet, however, were such that I could get on to it unseen.

In the twilight before dawn on 23 May, I rowed quietly over to the hump of rock and pines which screened me from all the nests, thanking the powers that be that no sentry gulls were on the outlying rocks so early, and cautiously hauled the boat into a deep cleft which hid it from all but a direct frontal or overhead view. Then I belly-crawled slowly over the rounded grey rock, into the thick heather below the dark pines, opened the hide flap and inch by inch, carefully and slowly, slid inside. Not one gull had seen me arrive or had cried out. Triumph! Then I looked through the camera hole.

From what little I could see most of the gulls were still asleep on their nests, but a few of their mates were stirring and here and there one or two walked a few steps, or took a symbolic peck at a nearby colleague, as if asserting the correct distance. The divers' area was wreathed in dark shadow but I could just make out a chunky streamlined form sitting on the nest. As I set up the camera with aching slowness, involuntarily my heart began to beat faster.

When the sun came up it was to light a view so magnificent no stage or film director could have conceived it, and I truly felt that somehow I had entered paradise. The diver sat motionless, her bright orange eye unblinking, her startling plumage so perfect she looked unreal, like a glazed precious ornament. Wavy black and white lines around and below her sooty black throat patch merged softly into the pinky-cream of her belly feathers. Symmetrical black and white griddles, shimmering like moonlight on ruffled waters, adorned her blue-black wings and back. The sun's rays burnished her shoulders into greeny gold highlights, and round white spots, all of different sizes, freckled the neat round edges to her outer wing feathers. Two rocks, blazing with diamonds of sparkling mica chips, led the eye into a profusion of 'kingcup' marsh marigolds behind her whose broad bright yellow flowers swayed iridescently in the light breeze, shining the early light back to the sun. Behind her the green stems of a patch of marsh grass thrust out tips that seemed afire.

Transfixed with the beauty of the scene, I dared not click the camera but waited each time for a gull to call out to mask the slight noise of the shutter. Occasionally she turned her moulded head with its dagger beak to look at a gull, up in the sky, and when she stared towards me her neck and head had the look and grace of a cobra.

After half an hour my neck ached and I lowered my head to rest on the soft heather. When I looked up again it was to see her sliding down the shallow runway towards the water. I had read many times that black-throated divers cannot walk; their strong legs and webbed feet, so ideal for swimming and diving after fish, are situated too far back on the body, so that they cannot balance. As she slid forward, thrusting herself along on her belly like a ponderous lizard with driving scrapes of her feet, it seemed true. I wondered if her mate had arrived to share incubation duties, for divers are believed to share the work – but no. As the sun was shining I supposed she knew the eggs would keep warm enough while she went to feed.

What surprised me was how the gulls tolerated her presence. Not once did I see them harry her, or while she was away, go near the two light brown eggs with their handsome black markings. I reached into my cloth bag to change the film, realising with chagrin I only had one colour print film left after working with eagles the day before. Well, it would have to do but I wouldn't waste any of it on the gulls. I was tired out too as I had just put in a thirty-four hour watch in the cramped hide up on the eagle cliff and, being roped to a rock, had had very little sleep.

I must have dozed for an hour but I was lucky to wake up when I did for suddenly the diver came paddling into the lagoon from the left, reached the shore, put her beak to the ground as the water still lapped round her, hauled herself almost upright in this fashion, then started to WALK up the slight slope to the nest. *Click!* And I had a perfect picture, as it turned out, of the seemingly impossible – a black-throated diver *walking*. She only went about three yards, stumping along until she had covered the shore rocks, then flopped herself down gratefully on to the grassy runway with a slight thump and shuffled with thrusts of her legs back up to the nest. There she stood on her 'hocks' as it were, and bending her head and neck down in graceful snakelike curves, delicately turned her eggs over with her long open beak, hooking them under the fluffiest feathers of her belly, then sat down on them again.

She seemed so close now; I could actually see her breathing and was astonished at the strange rhythms. She breathed in slowly eight times, her body swelling greatly, then was still for 1¼ minutes; then another six slow breaths and a long pause of about the same time. After that came six more breaths with an 85-second pause, followed by another eight. Clearly divers can store oxygen. Not only have they evolved this system of breathing for their long underwater dives but it has become their normal above water system too.

After an hour and a half more, I heard a low guttural call and there, paddling into the lagoon, was her mate – with his head lowered along the surface as if begging her to join him. He did not come out to take over the eggs. Again she took two awkward steps (it was harder to walk downhill), flopped down and scuffled out to the water. As she swam up to her mate he raised himself and showed off his cream belly, arching his wings slightly, his beak straight up in the air as if stretching himself. They did not touch each other. Then she dived with a looping movement, closely followed by the male and they were lost to my view.

I must have dozed off again for the sun was sinking when I woke and I missed her second walk up for she was already back on the eggs. I took several more shots, waited until I was sure her mate was not in the area, then slid out of the hide and down to the boat and silently left the islet. Next day, after ensuring from the eastern bay that one of the birds was on the nest, I boated out to make the 88-mile return drive to the inland town to buy more film. I also received urgent revisions in the post, for at last my Canadian wilderness book had been accepted. With these to complete, plus the eagle work, I had no more time to watch the nesting divers again that year.

On 30 May I was boating back from checking an owl's nest in a hollow tree to the east and made a wide detour of the islet in case the divers were still on the nest. I needn't have bothered. About half a mile beyond I saw the unmistakable form of the two divers. One of them had an odd bump on its back. I looked through the fieldglass – it was a downy dark brown chick, hitching a ride on mother's broad rear! As I watched it slid off and paddled between its parents; diver chicks enter the water within hours of hatching. I wondered what had happened to the other egg, but decided not to visit the islet for two days in case one of the pair was still incubating it. Also as the first fishing boats of the year were puttering past, I didn't want to draw attention to the nest.

But on 1 June, when there had been no sign of the birds since, I rowed over. The second egg was still there but cold as ice. It felt strangely light when I picked it up and as I shook it gently, the few contents swished around inside. Probably infertile, it had clearly become addled and when I sniffed the egg, it stank. I wondered then how the divers knew it wouldn't hatch. By its lightness, because they could feel no chick movements inside the shell, or could they scent it too? Most birds, it is believed, have no sense of smell, but perhaps divers have. After all, they can dive to seventy feet or more and must often have to hunt fish in dark water or at night. Maybe by some intricate taste or smell device, by filtering water through their beaks, they can trace fish. It is a theory that remains to be tested scientifically.

One phenomenon in the divers' world, which I have been lucky enough to witness for several years, occurs in late August. One 23

August, a hot, hazy and sultry day, I heard strange goose-like sounds, a grinding honking noise with an occasional high wail, though the wails were much shorter than usual – more like a very loud metallic double curlew cry, '*weeoo whirlyee, weeoo whirlyee*', ringing across the loch. I rushed out with my tape recorder, secured the last sequence of sounds, then had to sneak back and out again with my fieldglass and camera, for the divers were heading for my shore beyond the leafy trees.

I stalked them in bush hat and camouflage jacket, and found it was a red-throated diver with her chick. The parent's light browny back, rusty throat patch and gently uptilted bill showed up clearly, while the youngster was much smaller and had a white front. As I clicked away from behind a tree, the mother dived for about twenty seconds then came up with a fish in her beak. Then she swam along the surface with her neck stretched out and, as the chick swam to her, she swished it sideways to and fro on the water then slid the fish over. The youngster also swept the fish round on the surface, twisted its tail outwards and swallowed it. It looked for all the world as if the mother was swishing the fish around to show the chick it was a fish, and that fishes move and swim, and have to be caught. The youngster had done the same perhaps to acknowledge that it understood. The parent then rose up in the water showing her white breast and flapped her wings hard before subsiding again.

As I clicked away, bemoaning the bad light which forced me to shoot at 1/60th of a second at f5.6 on the 640 mm lens, I looked up to see the black-throated diver with *her* chick calmly swimming past westwards, a mere forty yards or so from the red-throats! Neither showed any animosity towards the other.

While black-throats prefer the largest fresh water lochs, the smaller red-throats nest on the smaller high dark hill lochans, sometimes as high as 2,000 feet, and both parents have to work as a complete team, sharing incubation and chick-rearing duties so that the one which is relieved can fly down to the preferred feeding habitat, the sea. This they have to do for the month it takes the eggs to hatch, and also for the two months more needed for the young to fly.

In the years since I have again seen red-throated divers and chicks on the loch, never before late August, and often not far from black-throat families. If it is interesting that the red-throats use the big loch as an intermediate stage before taking their youngsters to the wide open sea, it is even more so that the black-throats tolerate such 'rivals' in their area, though of course after the nesting weeks the territorial instinct would be less. Anyway, red-throats are a different species. What part territoriality plays in the rarity of black-throated divers is hard to know. Certainly the pairs that I have seen seem to put at least three miles between each other in the nesting season. Possibly the amount of undisturbed water is the main factor.

In the sixth season it was on 30 March that I first heard the cat-like mews of the divers, three weeks earlier than the year before, despite the coldest if driest winter in many years. A minute later came the harsh gobbling '*powoo powlyow*' calls too. I was still not totally convinced these sounds weren't made by the mergansers. Five days later I had conclusive proof, for I saw the pair making them as they chugged west in the centre of the loch. When they mewed they stretched their necks out on the water and back again, then arched their heads as if looking into the depths.

They did not nest on the islet that year and I only caught glimpses of them in the summer. On 30 July I looked up from my desk to see what looked like big ducks swimming along way out past the islet. I put the fieldglass on them – then rushed for the camera. It was a party of *four* black-throated divers – two parents and two chicks sandwiched protectively between them. The young, just about up to the nine-week flying stage, were ashy brown on top, with whitish underparts all the way down from the beak, roughly the same drabber colours the parents would have in winter plumage. What a grand sight!

In the seventh season I heard the divers' harsh courting calls from far across the loch in May but didn't see them near the islet at all. On 16 June I was boating west to work on eagles when I saw a pair, again with two chicks, swimming near the sandy point three miles from my home. This, again, was a tremendous sight. One RSPB study a few years ago revealed that the rate of success for black-throated diver nests in Scotland could be as low as 5 per cent – all the rest failing for a variety of reasons, mostly human-caused. Here, to my knowledge this pair had reared seven chicks in seven years, though I had no way of knowing how many survived to breeding age.

My heart still skips a beat when I first see these magnificent creatures arrive. Here, like the eagle or the wildcat, is a creature beyond man, rare because it is beautiful, rare because its needs are sensitive, rare because it requires solitude and has for centuries survived in true wildness. It is the true spirit of the magical loch. And as I look yet again over the shimmering waters and ponder their unsolved mysteries, knowing I have barely scraped the surface of their wonder, I will not temper my enthusiasm with dismay because of human shortcomings.

The loch will outlive me and possibly all human life and all I can do is seek a little longer to know its heart.

PART FOUR

Chapter 11

The Woodlands in Summer

Tired after a trek in our first summer, I went to sit between the buttress roots of a huge oak in the woodlands near my home. Everything rested in a drowsy hush and I felt I was in a vast natural cathedral fashioned by a mighty architect. Overhead the dark oak leaves shaded the brightest light while nearby the feathery spires of the giant spruce combed the sky. Not far away, from roots clutching the earth like great talons, the smooth green-grey trunks of beech trees reared gracefully like the pillars of unending knaves. Their slender branches held out horizontal triangular plates of small emerald leaves and through this dense mosaic of green, sunlight seemed rendered ancient as it filtered to the forest floor. The primroses, white wood anemones and golden celandines of spring had gone now but the slanting rays lit up swaying purple foxgloves and a last wilting bluebell or two as if through stained glass windows. It seemed almost holy, a natural sanctuary in which to rest amid the quiet splendours of nature. But as I sat silently in my pew of roots, my arm round dozing Moobli, life began to stir.

A rustling nearby, shrill squeaks, and the bulbous nose of a shrew quivered about the carpet of old leaves. Then it ploughed through the forest litter like a tiny bulldozer, searching for insects to satisfy its voracious appetite, and vanished behind a small mossy rock. The shrew has to eat both day and night for it has the highest metabolism of any mammal and can die of starvation after two hours without food. If it's lucky and lives as long, it dies of old age at eighteen months, and in that time may have reared up to three families.

Suddenly, high to the left, sounded excited '*pink pink*' calls of chaffinches which were mobbing a tawny owl roosting among ivy in the crotch of an old alder on the banks of the burn. They had a nest in a rhododendron nearby and were trying to drive the owl away, not only warning other birds but also teaching their young to recognise an enemy. As great tits joined the fray with scolding scissoring *chirs,* the owl shot away with unusual speed for a bird that normally floats along silently. Perhaps it would return in the dark to prey upon these very birds when superior night sight gave it the advantage.

In a nearby clearing, a blaze of colour where blue speedwell flowers, tall nodding foxgloves, wild violets, dog roses and white wood garlic grew among tall grasses and cream-flowered bramble bushes, butterflies flitted, seeking to mate in the haunts of their caterpillar youth. Small tortoiseshells and glowing peacocks with an azure eye upon each wing were the brightest, their flashing colours enabling them to locate each other at a distance by sight. Green hairstreaks skipped about like chips of jade and speckled woods hopped and flopped precariously through the air in the shade of the giant trunks.

I felt a light touch as a little thick-bodied brown butterfly landed on my hand. I saw tiny orange-yellow 'plates' decorating its wing and was astonished to recognise a rare checkered skipper. I knew it to be verging on extinction, confined to small colonies in south-east England, with a few round the Fort William area. The caterpillar of this rarity feeds on heavy seeded *bromus*-type grasses; by sewing the edges of grass blades inwards with silk, it makes a little tube from which it feeds shyly at dusk. It spends winter in more luxurious quarters – sewing two blades together.

When I saw this little speck of flying sunshine on my hand, I felt here was even more reason to scythe the bracken from the front pasture, and to encourage and plant its food grasses.

Suddenly Moobli stiffened and but for my restraining hand would have got up. A little roe deer doe I had not seen, browsing on brambles at the clearing's edge, scented us and quickly bounded away, followed by knobbly-kneed twin fawns, her blazing white rear tush floating between the trees like a lantern held by a running ghost.

As the sinking sun began to wink through the trees, I noticed a vine of honeysuckle had caused an ash sapling to grow in a spiral – a fancy walking stick one day – but as the long pink and yellow trumpets now released scent, a rare humming bird hawk moth flashed in from nowhere. Hovering, its wings a blur of dark grey and orange, its tufted black and white tail spread out to give quivering balance, it uncoiled a long proboscis and carefully inserted it into one of the flowers for nectar. Then a bumble bee burbled clumsily among the flowers, failed to get its huge body down a few and finally landed on a more open trumpet, thrust in a thick tongue of surprising length and got its reward. I decided to encourage honeysuckle bushes wherever I found them.

It was time to leave. We swung east along the main burn. Moobli soon scented something and led me to a low muddy patch left after the torrents of spring had died. There the five-toed tracks of a badger, with long serrated ball prints, showed clearly. Last night it must have left its sett and dug between the roots of a large tree a quarter of a mile away. Some shallow scrapes showed it had come to the wood to dig for grubs, beetles, and its main diet – worms. Back in the wood a leaf moved near my foot: two black and orange striped sexton beetles were digging frantically to cover the body of a dead shrew. In it they would lay their eggs, then stay by the carcass to feed their young until they were big enough to help themselves. Such beetles are the useful garbage disposers of the woodland floor.

The little pipistrelle bats which lived in my woodshed roof were now whinnering around the glades on skinny wings, feasting on the dusk-flying insects. An orange drinker moth hurtled across in zig-zag flight. One bat, homing on to the echoes from the moth's body of its own sonar squeaks, went in pursuit. Up-down, up, down, the bat a fraction behind, until the large moth shot through a screen of leaves and disappeared. While I knew the big black old lady moth could escape by suddenly closing its wings and dropping to the ground, this was the first time I had ever seen a moth actually outfly a bat.

It was then I realised how wrong I had been to regard the woods as merely a quiet cathedral-like sanctuary, where all seemed peaceful and things happened too slowly to be really felt by a human. Although I had known from boyhood in Sussex woods, the long treks in Canada and the first years in the Highlands that varied woods gave the finest habitat for wildlife, I had never really understood them as a complete entity. That afternoon, lucky to have seen life operating on several different levels in a mere hour, the vision of woodland as a *whole* came clear to me. No mere cathedral of nature, a wood is a great living *theatre,* and unlike a meadow, marsh or mountain slope, a theatre with many stages.

Stage One is the rock and subsoil in which badgers excavate their dens and trees and plants have their roots.

Stage Two is the woodland floor of soil and litter on which every living creature in the woods depends, where the food web of nature really begins. Here bacteria, so tiny a billion would barely cover a finger nail, so prolific a single cell breaking in half ad infinitum can produce 16,000,000 progeny in twenty-four hours, are helped by a great army of mites, springtails, eel worms, woodlice, centipedes, millipedes, earwigs, spiders, harvestmen, beetles and earthworms to break down the fallen trees, leaves and dead plants into the rich humus all plants need to grow. Old stumps and dead trees give footholds for mosses, lichens and fungi which also help the breakdown process.

Upon this soft yielding stage come a vast procession of characters to feed on the plants and smaller creatures – predatory insects, voles, shrews and woodmice, red squirrels, hedgehogs, frogs near the burn, toads, and daytime birds like finches, blackbirds, thrushes, tits and wrens. Some of these become food, too, for the 'heavies' of the woodland cast – the night prowlers, free from competition with daytime creatures, safe from man and his guns. Years of tracking show me how the fox slinks along the lochside, up the banks of the burn, then back round the edge of the wood, leaving its news scent at 'recognition posts' for other foxes. Then it works inwards as stealthy as a cat, preying on an unwary vole feeding in the grassy clearings, locating a low roosting bird by its excellent sense of smell, snapping up insects, beetles and even worms when hunting is poor. The wildcat hunts silently, pouncing with devastating speed upon small prey, and the badger snuffles along, ready to dig up young voles or mice from their burrows but happy with beetles, various bulbs, and its favourite earthworms. It, like the fox but not the wildcat, stoat or weasel, is seldom fast enough to catch the woodmice for they can climb like monkeys and also use the next of the woodland stages, to feed upon berries, nuts and fruits.

Stage Three is the area of plants and small bushes where many insect larvae feed and small birds have their nests. Here too come the roe deer – and in winter the red deer – to browse upon the plants, leaves and twigs of bushes or the succulent tips of unprotected trees, or even nibble bravely at the lower prickly leaves of holly. I soon realised it was foolish forestry to remove such 'weed' bushes for in winter hungry deer are then forced to strip bark from young trees and eat all the regenerating seedlings they can find.

Stage Four is the understorey of the trees where many more larvae feed and birds like goldcrests, tits and finches help themselves to both them and young buds and autumn seeds and also build their nests.

Stage Five is the high canopy of the trees, used mainly by insect larvae, small birds feeding on buds, seeds and the larvae, and for nesting by larger birds like crows, herons, buzzards and hawks.

Stage Six is the space in the glades between the trees and one of the most interesting of all. Here are held the aerial ballets of many insects who prefer the quiet air near their food plants to the winds of the open fields. Here too come the night-flying moths who help bees and flies to pollinate the flowering bushes and plants which they locate with scent-sensitive antennae when daytime butterflies are at rest. Upon them all at dusk the bats come to feast – the long-eared bats stealing a march over the tiny pipistrelles by plucking insects directly from the high foliage.

Lord of the night by far is the tawny owl. It works all the woodland stages from top foliage to forest floor. Locating prey by sound and through the pupils of its highly sensitive forward-facing stereoscopic eyes, it flies silently on broad soft wings, snatching climbing woodmice and birds from tree roosts, or sinking its talons into voles, shrews, or young rats on the ground, who often die without ever seeing the gliding shadow that hit them.

By day great spotted woodpeckers undulate through the glades with flashes of red, white and black, sometimes pausing to stab into an ants' nest before flying to the big trunks to drill for wood wasp, beetle and moth larvae. Often they share the trunk with secretive tree creepers who work spirally up from the boles, searching every cranny with sharp eyes for tiny insects. Once, when I actually had the camera in my hands, one landed on a larch tree right next to me and I hastily took a photo. Too hastily, for it was the last frame of a film, and a second later the tree creeper pulled a long white grub from the bark, paused with it squirming in its long curved beak (which would have been a better picture) and then flew off.

But for birds, insect numbers would soar out of control in the summer; nature's restraints are always at work. The birds hunt harder than usual at this time for they have young to feed. Yet they too have their controllers – like the dashing cavalier sparrowhawks which never take more than a fraction of the surplus birds, and then usually the less fit or wary.

Over the years I have learned it is upon these stages, through the pageantry of all four seasons, the interplay of day and night, that nature plays all the mysteries, dramas, comedies and tragedies of creation itself. Often I go, admission free, a humbled audience of one, to watch and wonder.

After seeing the sparrowhawk pair cruise by the bird table a few times in the first winter, when the male hit Corporal chaffinch, I wondered if they would nest in the nearby woods. But in the next spring and summer I only saw the female once. I was taking cabbage thinnings from my garden for salads on 21 July when she glided close by and swooped up on to the branch of an oak. I looked but couldn't find any nest.

Both hawks cruised the garden the following winter. I was in the west wood with Moobli on 8 March, standing behind a thick holly bush, when I witnessed one of the female's hunting methods. She had spied some bullfinches and chaffinches de-budding an old cherry tree near the wood's edge and was working her way towards them with immense cunning. Gliding low, keeping the trunks between herself and her prey so she was mostly out of sight, she descended, landing twice on branches with the least movement as if planning how to cross the next space. Then, when she judged herself close enough, she made a sudden dash and with an audible thwack struck the uppermost cock bullfinch with unerring skill, seeming to go for the main blood vessels under the wing because as she bore it away in triumph it was upside down.

Sorry as I was to see the demise of the beautiful little bird, the sparrowhawk attack at least kept the rest of the raiding tribe out of my fruit trees for a few days. In early April, three days after she had hit Pete chaffinch off the bird table and bust his leg, I was watching one of my wildcats when she made a sudden swoop towards the young cat, abruptly changed her mind, swerved violently and shot right out of the wood.

I was sure she wouldn't nest in the little wood after that; she had the trees of a two-mile-long wood to the west as well as lochside woods to the east to choose from – and on 5 April I had even more reason to believe this. It was the day I saw the cheeky blue tit loop up after a kestrel, and in the afternoon I actually saw the kestrel fly from an old crow's nest fifty feet up in the end branch of a leaning rowan tree on the north side of the wood. It seemed likely she would lay eggs there. Pleased as I was by this possibility I recalled the old falconry saying, 'A sparrowhawk for a priest, a kestrel for a knave.' It seemed all these birds of prey had made an unconscious judgment of their landlord anyway. And it was unlikely the hawk would nest so near the falcon.

Twelve days later Moobli sniffed the air, then went below a triangle of crossed fallen trees in the middle of the wood. The area was filled with splashes of white droppings – their scent had attracted him. I looked up and at that moment, from a new nest in the top of a tall Scots pine, the female sparrowhawk flew. It was seventy feet high and none of the trees near either nest had many branches to that height, so photography from a tree hide was impossible for a lone human.

I was astonished that kestrels and sparrowhawks would nest within fifty yards of each other. Later I worked it out. Kestrels are open field hunters; their ability to hover with deathly stillness in the sky makes it easy for them to spot small mammal and invertebrate prey below. Sparrowhawks on the other hand, with their fast flight, mainly hunt small birds in woodland glades, their short blunt wings and long tails making it easy for them to swerve and dodge round trees and bushes. The birds occupy different ecological niches.

On 20 May, the last day of filming for television, both kestrels flew across the garden to their nest. Eleven days later I was in the wood trying to photograph a cuckoo with a low tone that was answering one with a higher pitched call in the east wood, when I heard loud squeaky chitterings from the kestrel nest. The little falcons now had young. Within a few days, judging by glimpses of white down on one edge of the nest, the hawks had chicks too. I kept out of the wood then, but on 11 June, trying to photograph a great spotted woodpecker flicking bark sideways off a Scots pine to find insects, I found a striped flight feather below the kestrels' nest.

That afternoon the male was hunting the ridges above the wood and he had one feather missing from each wing. Had kestrels evolved some mechanism for moulting *evenly*, so that they could still fly well? The white splashes below both nests had increased, and once I heard one of the sparrowhawks calling '*kik kik kik*' in the trees above, like a questing tawny owl, but weaker. Nine days later there were two more flight feathers, one primary, one secondary, below the kestrels' nest; the male now had three missing from his left wing but still only one from his right. It did not appear to affect his flight. My 'even moult' theory was evidently mistaken.

On 29 June I was walking round the edge of the wood with Moobli threading through the carpets of yellow wood cow wheat, when we surprised the male sparrowhawk on his 'plucking post' with his head turned the other way. He was less than five yards off but as I had only gone to look for edible fungi I had no camera with me. Then he turned, saw us, let go his prey and shot into the hill slopes above, landing on a dwarf oak. I looked at the plucked carcass – a chaffinch with its head torn off. The 'post' was an upended larch base, full of peaty soil with roots sticking out starkly, and it was covered with small feathers blowing in the wind, looking like grey fur from a distance. While the hawk now took most of his prey to the nest, he had been enjoying this meal on his own. I took the camera back to the wood several times after that. One of the young hawks made a laboured flight to an ash tree fronting the lake on 12 July, but before I could get the long lens assembled it flew back to the wood. Once I saw a flash of copper in a hazel bush, and automatically trained the camera on it. The resulting photograph showed the male kestrel tugging at a decapitated mouse between his talons.

There seemed to be only two young hawks flying. During the next fortnight they kept high, learning to fly in the tree canopy, fed by the adults there, before venturing closer to the ground. While I was fishing half a mile away on 26 July, I saw a young sparrowhawk fly to a nearby rock, stare at its reflection in the water for half a minute and then lazily cross to some trees on the far side of the loch.

I had seen nothing of the young kestrels, though two days later there were three falcons hunting the high ridges behind the cottage. One, redder than the others with more perfect plumage, though it had a feather missing from its right wing, I took to be the female. She appeared to be teaching two surviving young to hunt, for while they circled she dropped and hung, hovering, dropped and hung like that twice more, then fell like a stone between the bracken. She emerged with something black in her claws, probably a beetle, flew to a mossy rock where the two young joined her, one being fed the beetle. What eyesight! A buzzard is estimated to see eight times better than a human, and can locate small prey from half a mile away. Smaller birds of prey have superb eyesight too, all adapted to their own distances.

I never saw the kestrels in the wood again but photographed the young hawks high in the west wood trees on 30 July. One adult spotted me and gave ringing '*kik kik kik*' alarm calls to the young. Nothing moving in the wood would escape those eyes. By 12 August the fledglings were still being fed in the trees but no longer roosted in the nest for now they were ranging further by day.

Just how far I found out six days later when I went with Moobli to collect golden sand excavated by badgers from a new sett in woods five miles away. It was a calm sunny afternoon, the loch and sky a panorama of blue and gold, and I rowed home. Moobli on his run along the shore became bored with my slow pace and decided to go ahead and swim out to meet me. He was huffing along with powerful lazy kicks, broad head well out of the water like a small grizzly, when one of the young hawks flew out low over him, seemed baffled as to what this odd creature was, then wheeled and landed on an alder.

I thought that would be the last of them but in early September both young hawks were again in the west wood. I had just released the first wildcat family, mother and her two kittens, and I was worried the hawks might fall victim – until I realised they were covering a far greater area, and were also too wary and sharp-eyed. I had a slight scare one day when I found a hawk's tail feather, still with a tuft of wax at its base. Then I discovered blackbird feathers nearby. It was likely that one of the young hawks had had a hard time overpowering a bird not much smaller than itself.

Not until the third spring was I shown how even an adult sparrowhawk can make an error of judgment. I was boating up the loch when I saw the male stalking a fat woodpigeon that was feeding below gorse bushes. The hawk glided from bush to bush, then shot out to stoop upon the pigeon which saw it coming and with a great clapping of wings, took off. Within a few yards the fast hawk hit it, a puff of feathers floated in the air, but the woodpigeon kept going, over the loch.

The hawk then seemed to realise he could not secure the pigeon if it came down in the water, that he hadn't the strength to arrest its flight as the pigeon was far stronger, and let go. Just then the hawk saw the boat coming, the pigeon, hurt, turned and flew back over the land, while the hawk dashed into cover of the bushes as we went past.

Both hawks and kestrels nested again in high trees in the west wood the following year and were all gone by 23 August. I found it surprising that, whether they were nesting or not, the general balance of bird life remained much the same. Each spring single pairs of robins nested in both east and west woods, the west wood pair once raising four young in upended larch roots right under the hawk's nest. Only in the fifth year, after a mild winter, did a third robin pair establish a nest and new territory on the north edge of the east wood. They also exploited the thin belt of trees that lined the main burn's gorge above the wood. Chaffinches, the hawks' main prey, were prolific anyway. The wrens also seemed little affected. These small, secretive birds mostly keep to bushes and are seldom found in hawks's nests.

Wren hens are the most liberated females in the small bird world for it is the male who has to do all the nest building – and then do his utmost, with song, flirted tail and shivering wings, to entice a female to be his wife and lay in one of them. One mid-April morning I watched a male building in a crevice in the root cluster of three larches that had fallen together. He carried in beakfuls of dry golden bracken, perched outside it, then sang his complicated rapid song – a wren sings for about six seconds at a time but enunciates over a hundred notes like a mini-machine gun – despite his full beak before flying into the crevice.

In summer the woods were darker, statelier, for the horizontal plates of dense beech leaves, now dark green, threw back the heat of the sun and the spreading oak branches and thick needles of the spruces and silver firs allowed little light to the forest floor. Only among the tall larches with their light green fuzz of needles, and among the small hazel bushes where here and there a rowan held its crown of leaves as tight as a *gypsy* broom, was the light kind enough to my camera. But the darkness of the glades held no worries for the creatures who worked the air space stages above them – the bats. When I first moved to Wildernesse I found some pipistrelles had colonised the old woodshed's double roof, but when my first two wildcats adopted the shed as their quarters, the bats – probably because of the cats' scent – vacated it. They still flew round the area, but where was their new home?

On 28 March the following spring I found out, and first became personally involved with them. I was up a ladder fitting heavy new eaves when two tiny mouse-like creatures tried to shuffle away and hide as I lifted up the iron roof. They moved sleepily, not like mice, and hauled themselves along with the little hooks on the front joint of each leathery wing. They had the little ferocious devil-like faces that made superstitious mediaeval folk believe they personified evil. I picked them up and one, slightly larger than the other, opened its mouth of miniature teeth which could not have pierced human skin and squeaked with high pitched terror. I felt it was a female. I looked at their fine rufous fur and named them Rusty and Dusty. When I put them in a foam-lined box they hung upside down from their feet and snuggled together, apparently for warmth.

I knew better than to put them in a warm place. Bats that wake up and fly on warm days in early spring are at risk, for they often don't find enough insect food to replace the energy lost, and die of starvation in later hibernation. So I set them in the cool workshop.

Each day I checked them after work on the roof, and when I saw the big one Dusty had moved I picked her up. Again she squeaked, a sound so high and thin it could not have been heard at more than a yard or two by the human ear. Wondering if she was hungry, I proffered a minuscule bit of mince with tweezers. To my surprise she chewed it all up but then would take no more. The other bat was so comatose it wouldn't even move, never mind eat.

On the fourth day it was so hot, sun beaming from a cloudless sky, the iron roof sheets cracked as they expanded. By mid-afternoon the cottage was hot inside and I took a look at my bats. Dusty was stirring and stretching each wing slightly, and I saw her sides were going in and out about seven times a minute as she breathed. I set the box on the sunny window sill outside and in ten minutes she was breathing about eighteen times a minute, her little sides ticking away. After an hour I looked again – her 'bellows' were now moving so fast I couldn't count the respiration rate.

I put her on my palm, stroking her soft fur with a finger. Her tiny black eyes opened (bats are only believed to see clearly for about an inch), I felt her heart pumping fast, then suddenly her wings went out flat. She jumped, by pushing the wings down on my hand, and took off. It was like a butterfly clasping its wings beneath its body, a soft gentle brushing, and she whisked away. She made two loop-the-loops as if to orientate herself, then made for the spruce glade, screened from the east breeze by the larch trees. Within seconds she was scooping up insects in her mouth. After a few minutes she took off for a sheltered dell between the oaks and I lost sight of her. Rusty flew in the same fashion before dusk next day – and I was glad to see other bats were also flying.

What perplexed me was how they got into the roof cavity, for I had left no holes. One summer evening I saw two bats emerge singly from under the lead cladding of the chimney. Out through a tiny gap of less than half an inch came a head, then with a quick flip it was through and airborne. I wondered, as I often saw them hunting up to 300 yards away, how they found their way *back* if unable to see far?

I knew that bats fly around locating insect prey and avoiding collisions with twigs by use of their unique sonar system of echo-location, through which their high squeaks bounce back from objects to the sensitive skin of their ears and wings – but how do they find their way home from such distances? Experiments have shown that small bats can return after being taken six miles away, and of a colony removed twelve miles some 44 per cent got back to the home roost.

Radio tracking individuals proved they flew directly back, apparently guided by some 'internal compass'. Watching my own bats fly back from 400 yards or so also proved it had to be more intricate than sense of direction. They didn't just fly back towards the cottage but went straight to the lead cladding gap, or to a tiny hole beneath the eaves, hovered briefly, shut their wings, then scuttled in like mice. Possibly the particular wall or roof throws back an echo they recognise, but to what distance? And how do they get to within that precise distance?

Just before their evening hunting flights in summer, the bats chattered in high pitched tones, a shade more bass than shrews. By early August, when the females were giving upside-down birth, their wranglings, probably over resting space, became noisy.

I later found they would readily fly in daylight, in summer winds, and even in pouring drizzle in autumn, as if unwilling to be deterred by the wet when they were trying to put on fat for winter hibernation.

In the unusually wet seventh summer, when many insects hatched late or were swiftly killed by the cold and rain, the bats' numbers were swiftly reduced. Fewer females bred successfully and some young were apparently abandoned. I occasionally found one dead by the cottage walls. One poor mite hung dead by its legs from below the gutter, as if it had been ejected by the colony. The older bats suffered too, for the next season no bats flew until 4 June, and I only saw three that whole summer.

One early September afternoon in the seventh year I came into the kitchen to find a bat scrambling over papers on the floor. The bat wriggled, squeaked shrilly and tried futilely to bite my fingers with its tiny teeth. I put it into the foam-lined box with some flies – which were gone three hours later. The meal seemed to have perked it up so I brought it into the warm study and put it on a propped-up slab of larch bark.

Instead of climbing the outer surface the bat worked its way round to the back, then climbed up the narrow gap between the bark and the wall. It didn't trust to sight and thrust out, one by one, a long quivering arm, hooking each single wing hook into the bark to haul itself forward, *feeling* the way delicately. Once atop the slab it had a fine time, scratching all over with its hind legs, occasionally flirting out a long leathery wing and combing underneath it as if it had fleas, twisting its ugly little head and biting into the fur like a dog.

I noticed then that hordes of insects were crowding at the window, especially midges, attracted by the paraffin lamp light. All I had to do to feed the bat while I studied him was to leave the window open. In came the flies and midges, zinging round the desk, some immolating themselves in the mantle. The midges were attracted by the white paper, sight playing a part in their location of prey from which to extract a mite of blood, and I watched them squatting on the surfaces, probing with the six tiny daggers of their mouths, trying to find blood. Suddenly there was that soft brushing sound and the bat had flipped off the bark and flown to my sweatered arm. It squeaked as I picked it up, so I threw it gently into the air.

Round the room it flew, dipping, diving, and soon decimated all the midges, flies and moths. Moobli, supine on the floor, watched the bat in horror, as if it had something to do with black magic, and whined to be let out. Then the bat landed on an iron candle bracket that hung from the roof. There it carried on scratching and flirting its wings, washed them with its tongue, even yawned. When I let in more flies it launched into further feasting, then went to roost on the bookcase.

I kept the bat in the room another day and night until I saw two others flying outside at dusk and let it go. It went off towards the west wood and, against the dark trees, I lost sight of it. I hoped the recovery would last and that like the others it would have time enough before the real cold set in to stock its tiny body for the coming winter.

All bats received protection in Britain in 1981 and it is now illegal to move a colony without government advice, but my bats had been safe for many years before that. Anything that mops up midges and mosquitoes as well as the bats do is welcome to stay, and I am proud of the bats in my belfry!

Chapter 12

The Woodlands in Autumn

As autumn approaches, the sun sinking imperceptibly lower each day, the warm air rises from the summer-heated land, cold air rushes in to replace it, and the first real gales begin. In mid-September there comes a chill to the air and the winds blow the first leaves from the alders along the shore; a few hazel and rowan leaves are next and their nuts and scarlet berries begin to fall. One day whole twigs with leaves are torn from the oldest ashes and then, towards the end of the month, the first hailstorms from the north sweep the hills and woods, physically dislodging leaves, nuts and berries from the trees.

But summer is not yet done and early October stages short revivals when the gloom clouds roll away, and for a few days the sun burns from skies at their deepest blue. As I hasten to collect more dry dead logs for winter firewood, the woods are again filled with bird song – great tits chinking, chaffinches chipping, wrens firing staccato salvos, robins tipping silver coins on to a tray, and even tree creepers adding their thin squeaky voices to the choir. In the canopy long tailed tits are forming the first winter flocks with coal tits, their shrill cries mingling with the zinging of goldcrests. As I lie in the sun for an hour or so of healthy rays, the air free at last from summer's bothersome midges, it seems the birds are no longer singing merely to advertise their presence to mates or rivals, or to defend or re-establish territories. Their breeding season is over – no nests, no wives, no territories really now – and they appear to be singing for sheer joy in the last of the warm sunshine, as if to recall the rapture of spring.

Soon these halcyon days are over, the cold winds, rain and hail return, darkness advances to later in the morning, comes earlier each afternoon, sap-making decreases in the chilled trunks, before the most spectacular scene change of the woodland theatre's entire year. Slowly, just one or two leaves at first, the trees change colour. The witch brooms of rowans turn orange, then crimson and then dark red. The hazel leaves go yellow and then become tipped with orange. The ashes pass from dark to light green, then turn fiery yellow and one October day when I come up the loch in the boat the oldest ash, the first to go, will look like a fire behind the cottage. The larch needles change from light green to deep gamboge and fall driftingly in the breezes like dry rain, each tree an individual pillar of fire springing from the ground. The jade tips of the birches turn golden and begin to fall, but the oaks and beeches hold out the longest, until the stronger November storms. Then the dark oak leaves turn slowly sere, to the khaki of their childhood and then to russet; while the beeches slowly change to orange and finally to a deep ruddy brown.

Now the trees stand in the full glory of their autumn coiffures; giant or dwarf, they are gems of infinite variety, each a masterpiece in its own right. From the hills above, when the sun shines between the clouds and the winds blow, the trees toss their crowns to and fro and the forests become a mass of multi-coloured flames.

Yet nature does not seek to perpetuate her latest beauty and slowly, as rain penetrates the leaves, frost ices their hold on the twigs and wind and hail break them off, the trees undress for the winter. Within a few more days most of the leaves, glowing embers of their summer glory, are upon the ground – drifting, tumbling, whirlpooling at the whims of the wind.

When the first chill comes and insects wane in numbers, the bats are first to leave the woodland stage. The latest I have seen several flying at once is 26 October. It had been an unusually warm day, despite a north wind. The sun had roused them by warming the roof, and three came out to try their luck. The cold evening air and lack of insects made them return after a few minutes to where the thick walls beneath the iron roof sheets had retained some heat. Twice I have been surprised to see a bat flying in late December, but always after a mild day or two had warmed the roof. Such mild spells also bring out swarms of fat winter gnats from their shelters – while these half-inch heroes neither bite not feed, they dance a dervishly aerial ballet in the lee of trees or in forest glades before pairing off to mate. Among them the rare bat who braves the cold weather swoops for a small but needed repast.

The woodland theatre hardly misses the exodus of the bats for the time of feasting has come for a whole host of other characters, among whom I am a minority of one. The hazel bushes are filled with nuts, hanging in their acidic green skirts, acorns stare down at the ground, tight as eggs glued into their tiny cups, limp phallic cones of the spruces and the nutty cones of the pines begin to fall with little thumps to the forest floor, while the cupules from the wind-pollinated female flowers of the beech trees have swollen into prickly four-sided husks, each holding two triangular oily nuts known as 'mast'. The rowan trees dangle flat-faced sprays of orange and scarlet berries, and vermilion hips with their itchy seeds, rich in vitamin C, have replaced the pink and white blooms of late summer on the wild rose bushes. The bramble thickets I have refrained from clearing also bear heavy loads of blackberries.

Now the woodland floor is rich in fungi – great chubby cep toadstools sprout beneath oaks, beeches and larches, growing to three inches thick and half-a-pound in weight in just three days. On the boles of dead ash snags I have left standing for woodpeckers clusters of appetising honey fungus rear light brown tessellated heads towards the sky. (They are delicious fried in butter.) Occasionally I find grey oyster fungus and a head or two of the olive green Horn of Plenty, which looks like an old-fashioned gramophone horn. Sometimes stout pure white funnel-shaped toadstools sprouting from fallen logs defy identification in my field guides. Throughout the woods the orange-yellow umbrellas of tricholoma fungi stud the floor from late summer onwards like imitation suns, and amid the conifer needles of the west wood the delicious golden chanterelle I have gathered since late July still grow their fluted flaming trumpets, as if blaring silent hymns to heaven only they can hear.

At first I knew nothing of edible fungi and I experimented in ways I do not recommend to anyone else. I noticed which fungi were eaten by slugs, nibbled by mice or had been upended and their radii chewed on mossy rocks by the red squirrels. Reckoning if they were good for mice and squirrels, who like hazel nuts, they probably wouldn't do me any harm either, I fried very small portions in butter and ate them – waiting for stomach pains, or even worse. But when I still felt fine after three hours I congratulated myself on the new food supply I had thus discovered!

I never gathered more than half the crop and learned to scrape away a bare patch of earth with my foot where they had grown and shake them over it so their spores would grow again.

I soon found I had competitors. In the first year, after my successful 'testing' of the first big ceps, which taste halfway between a mushroom and an oyster, I hurried back to a new quartet of huge heads – to find they had vanished. Deer tracks nearby gave clues to the culprit but also helped confirm the fungus was not poisonous. I found later that badgers also prize these fungi and love even more the golden chanterelle. I once hung a hurricane lamp out all night to keep deer and badgers off one patch – until I realised I should only take a small crop and they were more needy than I.

Attempts to dry the big ceps whole for winter failed, for if I left them more than two days I found them riddled with unseen maggots that had eaten them into a gooey mess from the inside. They were also too big and just turned black and watery in the oven. I learned to pick them only when young, then to shred them into very thin slices to dry in sunshine. This applies to most edible toadstools but honey fungus, if gathered in a fine spell, can be dried whole. Most fungi send down long threads which rot old wood and help the army of tiny creatures to disintegrate it to humus. Honey fungus threads glow with a luminous blue-green light, and in the days when the only link between villages was a footpath, they were used to show the route at night.

In autumn many creatures hurry to the free woodland canteen. Small birds like finches, tits, robins and an occasional migrant brambling, are joined by mice and badgers who feast on the nuts, acorns, mast and some fungi, and even a fox or two comes along to nip off a few blackberries, the dark purple evidence showing up in their droppings. A constant hammering comes from the rhododendron bushes, not from woodpeckers but from great tits trying to persuade hazel nuts to deliver up their kernels. Now the red squirrel pair come bounding through the ash and alder trees that line the shore, and work all day to eat and gather food to take back to their drey in the west wood – to keep them going until the larch, spruce and cherished pine seeds ripen later. They also shred bark from the honeysuckle bushes which, along with moss, leaves and grass, lines their winter home.

In early October flocks of redwings, up to two hundred strong, come winging in from Scandinavia. They fill the woods with sweet '*quip quip*' calls for two or three days, raid the mast, hover with bright flashes of their red underwing patches to strip the rowans of berries and then, stomachs full, depart for the rich pickings of the farmlands to the west before continuing southwards. Powerful fliers for small birds, they seldom land for a rest on ships or oil rigs on their 400-mile flight over the North Sea, and after a day's rest they easily outmanoeuvre the resident thrushes and blackbirds when they hunt for snails, worms, beetles and caterpillars.

Redwings are increasingly liking Scotland, began nesting there in 1953 and today there can be upwards of fifty breeding pairs a year. I look forward to their annual appearance. Once I was lucky enough to get a fine photo of one hovering, a bright berry in its beak, at ninety yards with a hand-held telephoto lens. The redwings prefer the lower woods, for within days of their departure each year I can find rowans at 1,000 feet filled with berries, totally untouched by the flocks.

When I first decided to cultivate the wild blackberries I made an interesting discovery: the berries that grew on brambles entangled in rhododendron bushes were always the first to be taken by birds. The thick evergreen leaves of the bushes provided protection, so they ate near their homes rather than from the brambles out in the open, where the sparrowhawks might get them.

I concentrated on the open brambles near the cottage to provide fruit for myself. I cut back their probing leaders to keep the bushes tight, trained the fruit-bearing stems with bamboo canes to grow sideways like vines, mulched and manured them and cut out the dead growth each winter. Noticing the wild berries were biggest when long dried grasses had been bent back over the lower sprays, I lightly distributed hay cut from the front pasture over the fruits. From a mere two pounds of skinny woody berries the first year, I was able to bottle a full 25 lbs from just five small bushes by the fifth – luscious, full inch-long berries at that – and still leave more than there had been before to the birds.

By this time the first autumn winds, hail, rain and cold have killed millions of insects, and as the small seed-bearing plants are flattened, animals and birds turn more to the nuts and fruits like wild rose hips. Often I find where the bright red hips have been carried under bushes and nibbled by mice, and tiny 'dining places' under the dry boles of hazel trees filled with their empty holed nuts. The only complete nuts on the woodland floor now are empty or rotten ones. How do mice and squirrels *know* they are inedible without biting into them? One November day I found out.

I had just opened the front door to go on a trek with Moobli when I abruptly hissed him back. There, bounding with short hops over the front pasture like a little monkey was the male red squirrel. Every so often he stopped, picked up something small in both paws, sat up on his haunches, shook it rapidly near his tufted right ear, then dropped it impatiently. Then he vanished into the west wood.

We went down and I picked up the objects. Empty hazel nuts! So that was how squirrels knew they were 'empties'. They shook them.

At night now the woods are filled with sharp '*whick*' calls of tawny owls and the mournful yet musical hooting of answering mates as the adults seek to re-establish territories and the young owls look for areas of their own. The air grows colder and more frequent northerly gales bring the first sleet. Even the rutting red deer stags and their small 'harems' of hinds come down to shelter overnight in the warmer glades. Sometimes two young stags will be in the west wood at the same time, each roaring, trying to filch a few hinds from the other. By daybreak they are gone again, wending up the steep slopes to graze on the last grasses and short heather.

By the second year I had an elementary woodland policy. I felled no growing trees, and left many standing dead snags for woodpeckers, tree creepers and other species. I only took firewood and lumber logs from the fallen trunks, bucksawing up those nearest the cottage first. I left many old oak logs alone for these give the best food for the larvae of many insects and invertebrate species which are at the start of the woodland food webs.

My main problem was that the woods were not regenerating naturally, largely because voles, sheep and deer ate the new seedlings. The obvious, and costly, answer was to fence the woods off completely. But on my long treks in the mountains, where thousands of acres have been fenced off in recent years by government and private forestry interests, I had often seen deer walking along heavily mulched paths by these fences, gazing forlornly at their former grazing ranges, to which they no longer had access.

Red deer had used my little woods for many years for winter shelter and I felt that if I cared about wildlife it would be wrong for me to deny them this refuge. So, I just fenced the top and bottom of the land and left most of the side access to the woods open. What was more, if it was true that varied woodland, with broadleaf fruit-bearing trees like sweet chestnut and oak among the conifers, was the best wildlife habitat, then it behove me to try to create it. It was autumn; sap making had all but ceased; now was the best time for planting.

On a few calm but damp days (to eliminate risk of uncontrolled fire) I cleared areas of fallen rotting trunks, extracted firewood and burned debris to get potash for the garden and tree planting areas. I left 'weed' trees and bushes like stunted pines, holly, scrub birch, rowan and brambles for both rubbing posts and browsing. Then I drove to a forestry nursery at Fochabers on the east coast for the first little army of sweet chestnut, oak, Canadian hemlock and Douglas firs.

Returning in the dark with the bundles of young trees, having a task to land the boat in a fair gale, I puddled them by torchlight, and over the next few days planted them throughout both woods. While professional foresters working on the bare straight earth furrows upturned by a bulldozer can plant several hundred baby conifers each a day, I was planting carefully in strategic individual spots. Two sweet chestnuts I hoped one day would form an archway on the lower path. The young broadleaf trees had large root clusters and each had to be separately dug in, manured and puddled. The ground was rough, full of rocks and roots, and it took three days to plant seventy young trees.

As I didn't want to fence the woods off, I laboriously made individual cages from stock fencing or wire netting and took a further week setting them round each young tree, with stakes to hold them down. These cages had to be enlarged in following years as the trees grew but as I finished the planting and caging that season I felt oddly satisfied, that I at least had done a little to match theory with action.

Of course, I lost some – a few oaks died, probably planted in places that were too dark, and in the sixth winter, an exceptionally severe one, when I had to go away for some weeks through a family bereavement, the red deer broke down the cages round the hemlocks and oaks that hadn't grown as fast as others and stripped them of leaves and twigs. Even so nearly half the trees survived, the sweet chestnuts, oddly, faring the best of all.

By late November most of the leaves have fallen. They lie thick on the forest floor, a last shelter before the harshest days for thousands of tiny creatures whose descendants will break them down. The fallen leaves also protect the few mast, hazel nuts and acorns not eaten by birds or mammals – the very seeds their own decomposition into rich humus will ultimately feed.

Now come the first gales of winter, the heaviest of all. At night the corrugated iron roof rattles, despite all my repairs, and there is a great din of hailstones on it. The loch rises and, in the morning, the winds are piling great waves against the boat, awash on its wooden runway. It takes three hours to bail it out against malignant waters, hands icy cold, remove the engine and haul it out of harm's way.

As I walk back the winds shriek down from the hills, scattering the last hazel nuts, mast, seeds and berries far and wide, perfecting the 'forestry' work done by the forgetful squirrels and the feeding flocks of autumn birds which scatter seeds in their droppings and spill many too.

In the woods the great trees which strive and struggle against each other throughout their lives, writhing their deep underground roots into each other's territories to be the first to suck the mineral sustenance from the earth, which with their foliage overshadow and block out the light from the slower growing, so that the weak and spindly die, now appear to the human eye to be visibly at war. The tall, creaking, groaning larches bang against each other with loud thumps, breaking off each other's branches. The oaks fence with their neighbours with the tips of their wide spanning boughs, and the storm tears off many other dead branches which crash to the forest floor.

The final leaves too, soaked and made heavier by recent rains, are being ripped away. The burns have swollen into raging torrents, tracing great white cataracts down the hills, and clearing away the debris of leaves, twigs and branches from their courses. As they cascade into the filling loch, the rushing waters and debris sweep away the shore-side weeds and algae born through summer and down it all sinks, to provide fertile feeding grounds for the water creatures on which live the young fish who seek the deeper, warmer parts of the loch in winter.

How inextricably intricate is the balance of nature, the way in which all in the natural world has grown to live with the seasons; at differing paces, to struggle, to thrive and finally submit. Only man has learned to create his own artificial city environments and overcome the worst of nature to his own benefit – though he has yet to overcome the worst in his own. He loses greatly, too, in no longer being a part, a close witness, of the pageantry of the seasons.

In a few days the storms die and the first snow of winter begins to fall, floating gently down from a dark and windless sky, and in the clearings it settles like a blanket to protect surviving seedlings a while longer from the mouths of hungry animals. Now the deer shelter in the woods on all cold nights and I go out with Moobli to help feed them sheep cakes and summer cut hay in the glades.

One morning a huge stag leaped to his feet from his conifer needle bed in the spruce grove and looked as if about to charge. With oaks on each side of the spruce I unfortunately had him cornered and didn't think a clout with my plastic water bucket would be much of a deterrent! After a few steps towards us, Moobli more than ready to have a go, he dodged to our left, bounded to the burn, soared over it with an easy fifteen-foot leap and after a few more steps turned round to look at us again. Like the hind and calf herds, he seemed to realise we meant no harm, that the huge Alsatian was not warlike, for the stag remained to watch us go on down to the loch.

Now that the deciduous trees have lost all their leaves they are like humans without clothes, their giant trunks revealing the results of their lives, their real character. The mighty oaks, sprung from tiny acorns, hold their great arms out straight and proudly, as if spurning the forces of gravity, while some beeches show their great vitality, their trunks split into two separate boles, each grown as large as a normal tree.

In the colder air the sap has almost ceased to flow in the chilled trunks. With rain coursing down them and dripping from their branches, there is now a sad beauty about the trees. They tower there, brooding quietly, ready at last for the long cold rest ahead.

Chapter 13

The Woodlands in Winter

All through the icy stillnesses of late December, January, February and March, the oaks, beeches and larches stand stark and bare, a deserted air about them as if their spirits have departed, so it is not easy to know the living from the dead. When the first breezes herald new gales, the branches shuffle reluctantly and crackings come from the ice in their bark.

When the strong winds strike, a few snap off and crash to the ground with loud thumps. For weeks the burn is frozen over, the waterfalls glare with ice in weirdly moulded shapes, and six-foot-long icicles hang like sepulchral stalactites from the myriad small ledges that line the gorge.

By day the woodland seems empty, more lonely, like a vast theatre where the doors have been left open and the roof removed – just a robin flicking over leaves, a wren searching crannies among mossy roots, or a thrush moving with silent sneaky runs, pausing to listen hopefully for the bristly hairs of a worm tubing through the earth.

Sometimes the air is filled with high ringing '*si si si*'s as a party of long tailed tits, winter flocking with coal and blue tits, flit through the trees. With their long three-inch tails, and tiny bodies even smaller than wrens', they look like flying crotchets escaped from nature's music sheet. The blue tits prefer the early budding oaks, while the others make for the greener shelter and late seed supply of the pines and conifers of the west wood. Occasionally, on a warmer day, one of the red squirrel pair forages fitfully for nuts hidden under leaves but mostly they keep to their warm drey and nearest stores in the tallest silver fir.

In early December a roebuck and his doe move into the woods to nibble bravely at low holly leaves or seek the few leaves still left on the brambles and honeysuckle. I went out one snowy morning in the first winter and was surprised to see the buck merely yards from the cottage. I sneaked back for the camera and got a photo of him with his head deep in the tall grasses. His soon-to-fall little six-point antlers were bobbing rhythmically like the twin batons of a conductor keeping time to some silent sylvan orchestra. Mostly I could see only his rear. Pictures of deer's backsides are little more elevating than those of humans, so to get his head up I tried a trick I use with red deer – I gave a slight whistle.

To my surprise, unlike a red deer, which will pause to look for the cause of the noise first, the roebuck shot away immediately. A lesson learned.

Next morning I peered through every window of the cottage before going out and saw twin lights amid the bushes – one yellow buff and kidney shaped, the other bigger, whiter and more like an inverted heart. They were the rear alarm rushes of the roes. So the buck was now back with his doe, on the snowy north hill behind.

No whistle this time as I crept out and stalked on my belly, so that clumps of broken old bracken obscured parts of their legs and bodies. How nimble and dainty they looked, setting their feet down as softly as doves, browsing side by side, their black snub noses and white lips moving among the herbage. Bambi-fawny creatures, they could have escaped from a Disney cartoon, and I saw the buck give a tender nibble on the doe's neck, her dark-lashed eyes closing briefly with bliss. As she munched away, her ears switched in independent semi-circles, like radar scanners. As she looked towards me, the big ears came forward together, and with her dark nose and broad-domed forehead her face looked like that of a giant long-eared mouse above the bush.

I kept still but a pair of hooded crows perched in a tall old rowan '*kaah*'ed raucously and the stalk was over. The two roes wheeled gently and melted away between the trees.

Later I found the roes' bed, still warm, through a sheltered tunnel in a rhododendron bush. Some days later I saw the doe through the kitchen window – calmly lying down and chewing the cud by a bramble bush, a mere four yards from the cottage. How glad I was then I hadn't raked all the brambles out or other weed bushes, otherwise the roes would have needed to strip bark and 'ring' some of the smaller trees.

In late November, when Moobli was a year old, we found some deep scrapes in the east wood where a badger had been making typical screw-twists for beetles, acorns or worms, and Moobli actually tracked it by scent across the pasture and into the west wood. There it had again dug scrapes below the pines. Coming back at a fair pace along the steep north hill I almost fell over the roe doe lying down in old bracken in a rock cleft.

She capered off quite slowly eastwards, then turned north and vanished over a ridge. Moobli didn't see her, being behind and below me, but he soon started tracking her, nose-high, following her exact path. Calling him off with a quick hiss, a natural less-carrying sound I used to control him, I was glad he appeared to have top bloodhound qualities.

Next November the doe came back for two days. On 13 December I let Moobli out first and he got a strong scent, bounced up and down twice in anxious query but knew well not to bark; then *three* roe deer downwind of us got our scent and bounded away. It was good to see the buck and doe now had a fawn with them and were still alive, a tight little family group.

When the first snow blizzards strike, red deer hinds, calves and yearlings often spend their days in or near the woods too, not just the nights. They stand forlornly, scratching for old mast and plants beneath the snow, look for tree top twigs and conifer leaves blown down by the wind and clip the tops of any little bush they

Moobli, the gentle giant Alsation. He helped me trek after wildlife in 300 square miles of glens and forests.

Wildernesse in autumn. Fetching mail and supplies entailed a 13-mile return trip in an open boat.

A cock chaffinch feeding a youngster on the bird table outside my study window

Often it is impossible to launch the boat for supplies

Above. Despite many meetings with otters, a good wild otter photograph eluded me for eight years until I saw an otter emerge from the sea with a crab in its mouth.

Left. The heron whose life I had saved as a fledgling after a summer storm had dislodged him from his nest

A golden eagle mother will spend many hours with its chick, though not so long when it's seven weeks old

An eaglet ready for its first flight on 16 July

Black-throated divers were believed to be unable to walk because their legs, used mainly for underwater swimming, are set too far back. In a rare picture, this diver proves the contrary as it walks up a slight slope to its nest.

I was perplexed by neatly flensed carcasses which had been dragged about by strong animals – until Moobli tracked them back to a badgers' sett, where the boar emerged after his nourishing feeds of carrion

The tawny owl works all the woodland stages, from top foliage to forest floor

Dainty though they look, roebuck fray young trees, eat the lead shoots and have to be controlled in enclosed woodlands

A migrant redwing from Scandinavia, a bright rowan berry in its beak

Kestrels and sparrowhawks nested in the west wood a mere fifty yards apart. Here the male kestrel eats a mouse.

The most tragic sight – an old stag with a rear foot wedged between rocks in an icy river, waiting for death

The real spirits of the woodlands are the red squirrels, difficult to photograph in the wild

A recently born red deer calf lying in a hollow in the ground

A healthy hind, with full red summer coat, and her six-week-old calf

A master stag taking up his place among the hinds and showing a proprietary attitude, though not yet roaring

Adult buzzard and chick at the nest

'Up close and personal' with a female kestrel I nursed through a wing injury

The big stag began to roar, with swelled neck and black mane thrust out

The stag moved about to keep his hinds together, standing often to roar and warn rivals away

After wallowing in thick black peat, looking black and fearsome, the stag turned to fight the young contender shortly before dusk

Before mating a stag and hind will stand close together and she may lick and nuzzle him. The actual mating is usually brief.

One false step and one could be away to a slithering painful death

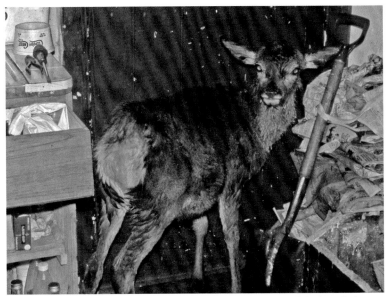

A weak calf was carried over several falls in the raging waters of the burn before I managed to rescue it and give it food in our kitchen

Death among animals is natural in the winter wilds, and the carcasses release vital nutrients to the soil. Around them the herbage is lusher, greener than elsewhere.

As she heard the click of the camera, a slim long-legged vixen jumped on to a boulder, intent suspicion in her dark orange eyes

A rare pine marten had moved into our woods, but it took several months to attract him to food on the bird table at night

can find on the lower slopes. Stags and hinds usually heft to different grounds outside the rutting season but after one New Year camp-out I was woken at dawn by a roar. We tracked about and finally found a six-year-old stag with the hind herd. This beast, an aberrant stag clearly, ran regularly with the hinds for three winters; once I saw him behaving as if in the rut, smelling the hinds and rubbing his chin over herbage.

During the sixth and coldest winter I woke to crunching sounds at dawn; two hinds were eating the ivy I had allowed to decorate part of the wall by the porch. When I went out to investigate I found they hadn't left a single leaf. I had always thought ivy to be poisonous to animals!

The woodlands in winter are an austere setting for tragedy. Occasionally some of the oldest hinds, heavily infested with any combination of warble or nasal bot flies, lungworms, liver flukes or keds, and weak from the cold and poor feeding, were unable to leave the woods in the morning and go up the slopes. With final pneumonia completing their troubles, they just died where they lay. A few young calves, mostly late-born, or whose mothers had been shot by stalkers, also succumbed.

One late December morning I found an old dead hind, her legs splayed sideways after a sudden collapse. Moobli scented out another between tangled branches of a fallen silver fir. She lay in deep mud without the strength to climb out, her head twisted upwards back on her neck and gazing pathetically at the sky, as if that was the last thing she wanted to see. She was very thin, her muzzle bare with age. Already it seemed rigor mortis was setting in because when I hauled her out her neck stayed stiffly in the same position. Her eyes were like huge hazel brown marbles and she had even lost the ability to blink. I got her to a drier more comfortable place then let her die in dignity for there was no way of saving her from irreversible decline. Later I skinned her and took the haunches as meat for Moobli, then hauled the carcass to an open spot to be disposed of by natural predators.

Occasionally the gatherings of crows and the guttural croaks of ravens revealed a new victim's last resting place. I studied these birds' methods, found they seldom came down to a carcass except to remove an eye until a few days had passed, as if to make sure there was no trap. Usually they ate in the near-dark of pre-dawn, for I seldom startled them from a carcass in daylight. The buzzards, like small eagles, ate latest and sometimes I saw them flap away, heavy with carrion.

Oddly, the sight and smell of the carcasses of their own kind appeared not to worry the live deer. They still came in at dusk and made their beds in the lee of rhododendron bushes, or on the mossy ledges of the thirty-foot-high rock escarpment in the west wood, sometimes just above the carcasses, whose heads were the last to lose hair.

When it snowed it was easier to trace the movements of other predators. Foxes seldom approached a carcass before four days had passed. Using my own eyes and Moobli's scenting ability, we learned to track foxes quite well. Out in a field beyond the wood where they could not walk round the snow – they will if they can – foxes left tell-tale straight lined tracks, even though further snow often filled them in slightly. A clear print under trees or on the muddy bank of the burn, showing the two front toes turned inwards, the ball smaller than a dog's, was another clue. A deer's stomach, hauled aside so the carcass wouldn't rot too fast, was the final one.

Sad though it appears to us, but for such carcasses (plus those of starved birds, mice, shrews, voles) many foxes, crows, ravens, inland cruising gulls and birds of prey would find it hard to survive harsh winters.

Most mortalities occurred between January and April, when both grazing and winter were at their worst. I was surprised when one 19 September we found a dead hind washed up under a fallen tree in the burn – until I saw her jaw was smashed, possibly by a poor shot. She hadn't been dead long for the keds were still crawling in her hair, looking for a new host. I heaved her into the wood, removed the haunches for Moobli, and what then happened with so early a carcass was interesting.

By 22 September she was covered with torpid bluebottles, all females laying eggs in the sudden bonanza of flesh, where they later turned pale and died. Three days on, masses of maggots were working the edges of the flesh. By the 28th the exposed areas were festooned with maggots, larger now, some falling away and striking off on their own, as if to pupate. I estimated there were some 4,000 maggots on the hind. Two days later they were flensing the skin off, working so frenziedly they had caused a foam to form. The fevered pulse for life as each struggled blindly for survival was quite eerie.

By 3 October all visible flesh, lips and tongue, had been eaten and the maggots moved inside the carcass. Three days later many came out again and covered the whole surface. By the 9th the horny black outer shells of the hind's cloven feet had peeled off. I checked again on the 13th to find the maggots' frenzy for life had been of no avail. The carcass had been reduced to bones and skin – scattered about by foxes, judging by Moobli's reaction to the scents. But for a mere dozen there were no maggots left at all. The foxes had eaten them too.

Of course, the deer are not the only sufferers in winter's depth. One December after a poor rowan berry crop and heavy snows, three redwing thrushes stayed in the woods. They seemed older birds, weakened by their long migration flight and unable to keep up with the main flocks on their southerly journey. They were turning over leaves for insects and beetles but looked weak, tottery, only keeping balance by sideways flicks of their wings, which they were holding low. Two weeks later I found one dead, and the remains of another – taken by a fox that had eaten it on a mossy rock, as if it wanted to see all round itself while eating.

The field voles were driven to desperate measures too. With grasses now gone and other herbage in short supply, they risked annihilation by my young wildcats in order to chew the deer cakes in the woodshed. Once, after a robin had hauled a piece of bread to the ground from the bird table, I saw a vole come wobbling out from a hole beneath the porch and start gnawing it. The robin, displaced by the vole, hopped about perplexed as the vole chewed away, decided the odd creature was not dangerous and flew back, beating its wings on the bread and driving the poor vole back to its burrow.

Each year the winter gales inflicted more damage upon the older trees. One dawn I was woken by a loud cracking above the howling of the wind and shakings of the porch roof, followed by a mighty thump from the woods. It shook the whole cottage. Worried what damage might have been caused, I dressed reluctantly and as the winds began to die down, went out with Moobli, who also seemed anxious.

We walked through the east wood and soon found the cause of the thump. A mighty beech, of such vitality it had grown twin boles a mere four feet from the ground, each one as big as a normal trunk, had lost its eastern bole. Ripped off by the gales, the entire fifty-foot length, itself divided into two trunks nearly two feet thick, had fallen across the burn, all the branches tangled between the rocks. I knew, from a previous faller, it would soon form a dam of other dead branches and trunks swept down from above, forcing the burn to course through the wood, destroying the sites of many fungi and springtime flowers, and the extra debris also sweeping away seedling trees.

Well, it was a sunny day and I was fed up with my desk. I got my new chain saw, waded into the burn and cut up the branches and twin trunks, stacking them in carryable lengths along the bank. It was hard tricky work amid the slippery rocks and as I cut the under branches from the main trunks they overbalanced and fell. I was lucky to leap out of the way in time. I cut up half the mass, leaving channels for the water to flow easily, then took Moobli for a trek.

Once, a tall larch which divided into a V near its top was blown down, but the V hit into another larch and the tree became hung up. In gales now the second larch swayed, lacerating its bark against its fallen brother. Such hung-up trees, called 'widow makers' in the logging forests of western Canada where I had once earned a living with a chain saw, are dangerous and cutting them down is risky. They have to be taken down in sections.

The method I had learned was to pile the saw into the back of the leaning tree at about five feet until just before the sagging weight causes the saw to be irretrievably pinched in the trunk, then cut upwards from the front until the second cut almost meets the first. There is a loud cracking, you leap away as the lower log falls, and the main trunk crashes down on its new base, 'walking' not more than a foot or two towards the standing tree with each successful cut. The V fork up top slides down the standing tree at the same irritatingly slow rate. It takes a couple of fraught hours thus to reduce a 'widow maker' and transform it into either firewood or building logs, which then have to be carried several hundred yards back to the woodshed.

In the third winter I again heard loud pistol-shot crackings and bangs. I rushed out to see that a freak whirlwind had come down from the hills and, while leaving the edge of the west wood untouched, was systematically destroying all the thick upper branches of a large oak. It was as if the tree had been selected by some wilderness god for individual destruction. I watched one branch twist as if wrenched by a giant invisible hand, then snap off as easily as dry spaghetti.

It was eerie, frightening, to stand in absolute calm only forty yards away and see a huge tree literally torn apart. As the whirlwind swept on down to the loch it lifted branches and tore off twigs from some hazel trees in its path, then erupted a small water spout from the surface of the loch before petering out. The oak was left looking like an old brown umbrella turned inside out, but it valiantly produces leaves each year and the thick foliage resting on the ground formed a favourite shady day-shelter for the wildcats in summer, and a new night bed for deer in late autumn.

There are times in the worst Highland winters, when the human libido has sunk to its lowest, that the isolated wilderness dweller feels not only that nature's cards are stacked too high against him but that some malevolent spirits have moved in, bidding him to get out and back to civilisation. An old tree crashes near the boat, a gale begins unexpectedly within minutes of launching it, the top of a silver fir breaks off and spears the ground two feet from where he walks in the woods.

There had not been a permanent year-round resident at Wildernesse for over sixty years and I was sometimes told by locals that the place was haunted. Every time I saw him one old keeper asked me the same question.

'Have you seen the ghosts yet?'

I got a bit fed up with this and when, on a rare visit to a local pub, he asked me yet again, I replied loudly.

'Yes I have. And I told them if they didn't quit bothering me I'd frighten them all to death!'

It was meant as a joke but no one laughed. I had under-estimated the power of superstition in the Highland mind.

I was told then that many years ago an old fisherman who used the place as a summer lodge had drowned off my south-eastern land spit, near the gulls' islet. The body had never been found. Shortly after this (it was before I had Moobli) I was walking through the woods of the spit when I heard two loud groans and a twig snap behind me. I felt an icy stab of fear, the hair rose on the back of my neck, but when I turned there was nothing there. Not until a month later did I find the cause of the loud groans. A small dying larch had been blown diagonally across the trunk of another and in certain winds the trees rubbed each other in such a way as to produce the remarkably human-like noise.

Once, after a really rough boat trip, I began having recurring nightmares about falling overboard and, hampered by thick clothes, rain suit and rubber boots, slowly drowning. I woke up from them bathed in sweat, resolved to buy the best life jacket I could find. In fact I never did more than keep the plugs fresh.

Oddly sinister experiences also occur with wild animals, which as time goes by, assume an ominous quality.

The most eerie time came in the January when my old tom wildcat Sylvesturr had mysteriously kicked out his heavy den door and, leaving meat behind, had forced an escape from his enclosure. Something must have scared him badly to make him act that way, for he was a former zoo animal, used to bars all his life. I found large but messed-up tracks in the snow near the pen, so it must have been something big. On the third day after I had set a large box-cage trap for him in the woods, the bait meat had disappeared, the heavy door had slammed down, but the trap was empty. Neither he nor any fox would have been big or strong enough to hold the door up while taking the meat. Moobli had been shut in the cottage all night.

Next day I was woken by the cronking of ravens. The carcass of a large hind had been washed down into the burn pool that supplied my water. I hauled it out to the central high, dry and hollow ledge which divided the waterfall into a fork, reckoning its eventual juices could nourish the bank of wood sorrel that grew in a broad cleft below.

Two days later the entire carcass had disappeared and though I searched the burn down to its mouth, thinking the rain-filled spates must have come high enough to dislodge it, it was not hooked up on the fallen beech remnants, nor anywhere else in the burn. When I searched the land on the far side I found a red deer carcass of about the same size thirty yards away amid deep brown bracken; it had been reduced recently to a bloody skeleton. If it *was* the same animal, one that I had been hard put to drag up to the hollow ledge, any beast that hauled it through the fast flowing waters would have had to be extremely strong. As I examined it Moobli moved about whining, cocked his leg on tufts as he did for foxes, but could not seem to follow any scent track.

Then in early February, when I was boating back from a supply trip, I saw a sinuous neck and head, with twin spikes on it, flopping in the water ahead. In the mist it looked like some legendary monster. It turned out to be a drowning hind which had tried but failed to swim across the loch. I skinned it for yet another rug and cut fresh meat off for Moobli, myself, and the wildcats. When I got home I hung the heavy wet five-foot-long skin across the thick logs of the porch windbreak to dry out. That night came a scary sequel to these odd events. I was sitting at my desk, Moobli dozing by my side, when there was a sudden loud banging and flapping sound in the open porch, as if the logs were falling down loose. I rushed out. The entire heavy skin of the hind had been physically hauled off the high logs on to the ground and the whole odoriferous tail section had been torn out. No fox or wildcat would have had the weight or strength to do that; a man would have found it hard. I began to feel fear then, and Moobli was oddly scared too, and showed no wish to chase after whatever had done it.

It was then I recalled a young naturalist with connections in high places, who had hiked in to see me some weeks before, telling me about some misguided 'conservationist' releasing two wolves into the hills of my area. Could it be true? A wolf certainly would have the strength to keep the trap's door open, haul the wet skin off the porch logs, and *two* wolves could have hauled the carcass from the ledge in the burn. But surely wolves running wild would have been seen before now, or their typical sheep kills found? There was an eerie atmosphere about Wildernesse and I felt apprehensive.

At the time I was also having correspondence with people in Caithness and Sutherland, where it was believed a cougar or European lynx was running wild, as dead sheep had been found with their skins neatly flensed off, in the manner of a large feline. From cougar experiences in Canada, I was working out on maps from sightings and kill positions where the animal might appear next.

A few days later I found a sheep carcass that had had its skin neatly flensed off too! In the long wood, over a mile from home, Moobli scented out the rear section of a large roe deer – but the rest of the body was nowhere to be found. Had some jokers released a *cougar* near me? I kept Moobli close by when I went out after that.

Then suddenly the mysterious incidents stopped. I retrapped old Sylvesturr, had to nurse him through a bout of pneumonia while rain, sleet and snow fell outside, and life returned to normal.

Not until some unusually harsh March weather did I discover the likely causes. I again found a sheep carcass that overnight had had its skin and wool neatly cut away as if by a pro skinner. A deer calf carcass I had hauled up into the hills for eagles had been a third demolished in the first night – certainly no fox's work. Despite bitter cold I kept watch with a torch – to see peachy-pink eyeshine from eyes too far apart to be foxes'. But the beam wasn't strong enough to distinguish any colours behind the eyes.

When I went up next day the carcass was almost completely consumed, spine severed in two places, rib bones chewed right down to the spine. Moobli scented about and twenty-five yards uphill found the final clues – rows of shallow toilet pits that could only have been dug by badgers, filled with loose oily black droppings. Moobli followed their line and I was surprised when he led me to an old sett we knew at a height of 600 feet over half a mile away. The sett had been unoccupied for a year but it too was now lined with deep pits filled with the same loose droppings, proving the badgers had overstuffed themselves with the pure meat.

It was my first real proof that, in dire winters when the ground is frozen and their favourite worms, grubs, beetles, acorns and underground tubers of wild plants cannot be dug out, badgers will readily eat carrion. I have since discovered too that a pair working together can easily drag a deer carcass many yards away into thick bracken, largely by repeated tugs on the meat and bones rather than from a desire to hide.

Well, I would get something out of the whole episode, I decided. I left the badgers to sleep it off for three days before going back with camera, flash gun and tripod – to get shots of the boar coming out at dusk. Big fellow he was too, weighing about thirty pounds. And in very fine shape indeed, his coat glossy after his nourishing feeds!

Another alarming incident occurred in late winter, one which not only ended tragically but which illustrated one of the human-caused problems from which red deer suffer. I was rudely awoken in the eerie pre-dawn twilight by the revving of a truck or car's engine. Although there was a rough private forestry track on the far shore, it was little used and at this time of the morning the sound made little sense.

Pulling on trews and sweater, I hurried barefoot over to the window and stared into the snowy gloom. There were headlights on the opposite shore – four of them. As I grabbed the fieldglass, they went out. Sweeping the glass in short arcs, I raked the area. Then I saw something in the water. It was moving out from the shore and seemed to be heading my way. Had they launched a boat? No one intending a friendly visit would set off at this time of day.

My heart beating fast, I put all my clothes on properly, set an axe and a crowbar by the front door, thinking they would not be much use against a gun or two, then with Moobli at my side I went out and took another look through the fieldglass. To my surprise both sets of lights went on again, engines started and both vehicles set off towards the east. But the object in the water was heading nearer.

I still could not see what it was. I hurried over the snow patches to where my boat was hauled up on the shore, then looked again. The object was almost halfway across now and suddenly my eyes could make out what it was – a swimming stag. Its antlers and head were well clear of the water but, as I watched, it turned in a short circle then stopped swimming. It appeared to see that the two vehicles had gone – their rear lights were vanishing round a point a mile away – and to start swimming back again to the shore it had just left. That water must have been near freezing.

I was sure then I had seen the vehicles of some poachers. They had probably seen the stag in their lights, and had meant to shoot it but it had foiled them by plunging into the loch and swimming out fast and far enough to make noisy rifle shots an unnecessary risk – for there would have been no way the men could have recovered the stag's body if they *had* hit it.

When I lifted the glass again I was surprised to see the stag had stopped swimming. Its head and antlers were flopping to and fro like a tree branch being manipulated from below, but weakly, as if it could swim no further. I ran back for rubber boots, the boat bung and my camera, got the boat into the water in record time and, hardly noticing Moobli prancing nearby, set off.

I roared out into the gloom but I could see the stag was now just an inert form in the water. By the time I reached him he was at his last gasp. A billow of water came out of his nostrils as I grabbed one antler to lift his head clear. He gave a spasmodic kick with one rear leg and then went still.

Hoping I could still save his life, I grabbed the other antler, took a deep breath and gave a mighty heave. I had him half out of the water and across the gunwale. I banged my right boot against the boat's side and with one more heave the rest of him, except for his back legs, slid into the boat.

Try as I might to pump water out of his lungs by artificial respiration, working his lower ribs in and out, none came out of either his nose or mouth. He was finished. I looked sadly down at the once powerful body; the terror of the drowning moments seemed to be still held in his wide open eyes. The heavings had cost me more than I had thought, for I found myself trembling slightly, and I sat down for a few seconds' rest on the seat.

Suddenly I heard huffing snorts nearby. Moobli, anxious to know what was happening, to be of help if he could, had swum all the way out in the icy water. I tried to lift him out too but my first attempt failed and he decided he'd be safer making it back on his own and set off for our home shore again.

I got back before he did, dragged the stag's body up the grassy bank to the first trees in our west wood and bled him after roping his rear legs high to a branch to assist blood flow. I had just started to skin him when Moobli arrived, shook himself fairly dry then lay on a nearby bank, panting as if he had just returned from a summer swim.

A sad incident but there was no point in wasting the meat. I cut the poor beast up and what I couldn't keep fresh for my own purposes I distributed among friends on my next supply trip.

At least the remainder of the stag's carcass did not seem to upset the other deer for they still used the wood at nights. Two evenings later when Moobli whined at my feet, I looked out to see the dusky forms of twenty-six hinds, yearlings and calves which had moved out of the woods into the front pasture. With their thick bodies and long necks buried in the grasses they looked as if they were hoovering down what winter herbage they could find.

One minor worry concerning deer remained however. For two years I had not seen the roebuck in the area, not even on treks or boat trips up the loch. The roe doe always appeared alone. Probably the large rear section we had found in the long wood had come from our buck, although I had no idea of the cause of death.

Slowly the dark days of winter pass, the long drizzly hours change to showers with glimpses of low sun gleaming palely through the clouds. By early January the daffodil bulbs have thrust their thick jade swords above the woodland leaf mould, and the first clustered lamb's-tails of hazel catkins are dangling, khaki yellow and fluffy, from the leafless trees. Before dawn in early February one mistle thrush sings lustily from the highest spray atop the giant spruce, while a rival tries to sing him down from the tallest silver fir in the west wood across the pasture.

However, winter is not over yet, and one bleak, cold and windy late March day I set off on a trek on the far side of the burn to see if the prolonged calls of a pair of ravens from a high cliff meant they were nesting there. We had just crossed the burn, carefully stepping from one slippery rock to another, hoping not to be drenched to the knees so soon, when I saw a distinctive white tush of hair a mere thirty yards ahead.

Checking Moobli back, I ducked down, crept forward and peered through some standing swathes of brown bracken. It was a roebuck, the biggest I had ever seen, weighing a good 50 lbs though its sides were pinched from poor winter feeding. The wind was ideally towards me but the light was poor and I took three photos as he grazed, his head too low to be seen. Having cast last year's antlers in early December, his new ones were just three-quarters grown and still covered in greeny-brown velvet, like little mossy branches.

Here were chances for perfect pictures of a wild roebuck but the light meter gave too low a reading for a 1/60th of a second exposure. 'Come on sun,' I whispered, and no sooner had I made the wish than the clouds rolled by, the sun gleamed down and up shot the meter needle to a safer shutter speed. I wished then it hadn't, for the buck was behind a hazel clump and the shadows of the slim trunks showed up all over his coat like prison bars. With Moobli obediently staying back (he knew his job by now), I jerk-crawled with one hand on the ground until I could see the roebuck clearly, but he would not lift his head. I whistled. He took no notice.

'Hey. Head up!' I said loudly.

Still he kept grazing. Was he deaf? Finally I stood up and walked towards him, focused, found I was now too near and actually had to step backwards again to get him all in the frame! Then he looked up, giving me some perfect shots as my film ran out. It was the easiest start to a photographic trek I had ever known. Even when I walked sideways down to the loch, the roebuck only moved away a few paces and stood watching. Moobli was breathing huffily, restraining himself with difficulty but enormous discipline.

Now the sun was beaming strongly, my head touched some hazel catkins and pollen flew out in a little cloud. We located the ravens' nest – when the cock bird flew in and called his mate off it – but it was up a high sheer cliff with nowhere to put a hide. As we came back I was delighted to see the roebuck had crossed the burn and was now in the lower end of the east wood. We made a wide detour to reach the cottage from the rear so as not to frighten him away.

Next morning, the first really sunny day for a month, I saw with a mixture of delight and regret some flashing white lights among the early creamy blossoms on my fruit trees, and heard faint '*dui*' calls. The lovely, cussed, bullfinch tribes were back again, devastating the buds of my future crops, and the 'lights' were the pure white rumps of the rosy-breasted males.

All summer, autumn and winter we see them not at all but come the first hint of spring, before the first daffodil blooms brave the late snowfalls, back they come, apparently from nowhere. Clearly, because ours are the only fruit trees in some twelve miles of lochside, they remember where they are and know the time to come. And somehow they pass that memory on to their children for after harsh winters the flock may be reduced to two browny females and three males.

Now they were hard at work, as acrobatic as tiny parrots, twisting and twirling and scattering blossom petals, five males and three females, a healthier nucleus. As we walked over, two of the males, dark eyes hidden in their black stormtrooper heads, flew up from the rainwater runnel where they had been slaking their thirst.

Sometimes I clap with my hands, sending them off into the west wood, in the hope they might get the idea to leave a *few* buds. But as I watched them today, then saw a little brown tree creeper lever up a birch in tiny spirals, pausing every so often to give a faint high-pitched song that always ended in '*tee wee*,' I didn't mind too much. I always have enough fruit left for my own needs anyway. I knew too that spring was really on its way, and I just hoped the new roebuck would meet the lonely doe.

Chapter 14

The Woodlands in Spring

Spring comes slowly to the wooded Highland world round Wildernesse, but as the earth turns in her year the sun's angle rises, dawn quickens earlier in the east, the days lengthen, shadows lift and the dark nights begin to lose time to the light. Northerly winds move to south-westerlies, bringing warmer rain and brighter clouds through whose gaps the first golden beams begin to shine, and the sap flows more strongly in the sleeping trunks of the trees. Rainbows form as showers drift away, to touch the woods or if I am on the shore to finger the cottage, and always if I advance, they move before me to highlight a rockface, a knoll, a lonely leafless rowan or spur of granite in the tawny hills above. Spring is nature's boast, but how gradually she bestows her renewal of beauty upon the wild. Like a great artist she adorns her living canvas, fills her stages and concert halls in infinitely varying ways, timing all her entries to the severity of the winter before; so that each year she seems to make a different entrance.

One year I notice spring's arrival in early March when I see my honeysuckle creepers have grown new leaves, and in the mornings are holding little cups to the sky, their wine the crystal dew. And when the sun retreats again next day, to allow winter to stage another frost, they become cocktail containers of icy globes. Another year she sends a wren as her second herald on 5 March, the first bird to mingle its strident voice with the clarion calls of the mistle thrush at dawn. Now I see a greenish tinge spreading over the twigs of the smaller ash trees on the shore. A few days later the larches are sprouting light green tufts, and a small flock of woodpigeons, trying to colonise the area, land in the female holly tree near my water pipe and feed greedily on the last red berries.

It is now that spring always strings two or three cloudless days together, the sun burning down from heaven's blue vault with pure heat upon the cold-gripped earth, and I keep my eyes – and ears – open for the courtship of the great buzzards. '*Kee-oo*' comes the high-pitched ringing cry above the silent hills, echoing between the woods, and being thus kindly warned of their coming, I dash for my camera to try to secure better pictures of these fine birds of prey flying.

I have never heard buzzards call in winter, and now it is most often the female which calls, keeping in touch with her smaller male companions, for in spring she often seems to be attended by two males. At first I thought the trio were the female, male and chick of the previous year. Over the years I have never seen two large birds with one small, which would happen when the chick would also be a female and as large as its mother. I have learned the chick is always independent by midwinter, that the two cock birds are courting competitors. Yet they never seem to fight each other, for it is up to the female to choose and she has usually done so by the end of March. Only once have I seen a courtship trio as late as 23 April – when possibly the previously ousted male had again come up on the pair in his search for a mate, though the true pair asserted their bond in front of him. One of the cocks made a strong display, raising wings high, flapping very slowly, and bringing them right down below his body each time as he gained height. Then he dived on to the hen, who turned sideways with talons out, as if expressing her preference. Next, the two birds made tight circles round each other, at times almost beak to beak, both calling loudly '*kee-oo*' . . . '*king-oo*' . . . '*kee-oo*' . . . '*king-oo*' and the discomfited suitor drifted away alone.

Some years there are only the two birds, the female with her smaller mate of the previous year, which have survived the winter together. Then the larger hen, her outer pinions spread wide like a small eagle's, stays higher in the sky as if possessed of superior long sight, while the male keeps lower, scanning the ground below rock faces. Occasionally he banks high, dives upon her and she turns on her side, almost touching talons, as if rehearsing for when they have chicks and he will thus present her with food to feed them.

Such sightings only occur on the few fine days for buzzards dislike soaring in the rain and do their hunting then from vantage perches in the shelter of trees or even telephone poles. One sunny day in the fifth spring I heard no calls but saw Moobli outside looking up at the sky. I went out and secured photos of both birds flying. I was surprised – as the morning sun reflecting off the loch lit them from below – at the sheer beauty and myriad colours of their underwings: whites, buffs, greys, browns and ruddy flames of orange feathers arranged in superbly intricate patterns. The buzzard was no mere mottled brown old bird after all.

It was in the sixth March I thought they might nest in the west wood for both birds often sat in a tall pine, warming up as the sun lit them into golden balls, and preening, the occasional feather drifting earthwards for my collection. One day the two-foot-long female cruised by close to the window like a dark shadow as I sat at my desk, her bulging dark brown eye staring in, as if ascertaining our whereabouts. But they did not nest in the wood that or any other year. I had established by then they were hunting at least seven miles of hills and woodland along the loch shore, and while buzzards do sometimes nest near human homes, doubtless with so much room this pair saw no reason to take unnecessary chances. I would find their nest one day . . .

The first fine days of early spring are alas short-lived and in March blustery winds and hailstorms return, killing thousands of newly emerged insects, telling nature winter is not yet done. At dusk during one such squall I was startled at my desk when a little long tailed tit, blown off course from the main flock, came scuffling at the window, as if trying to get through to the warmth and light. As I rushed out with a torch to help it, I saw it recover and fly into the evergreen leaves of the rhododendron near the path where it would shelter for the night. Next day spring entered truly when I woke to a calm hot day to hear the woods filled with bird song, the loudest dawn chorus I had heard at Wildernesse.

Each spring I tried to record which flowers came first, which trees first sprouted leaves, but eventually realised I would only end up with generalisations. Plants are almost as individual as people and much depends on the harshness of the preceding winter, the patterns of rain, wind, hail and sun of early spring, the ages of trees, and the amount of shelter, shade and light each tree or flower has.

First to flower on the woodland stage are the wild daffodils. One by one from mid-February like players entering an orchestra pit they raise their yellow trumpets to the chilly sky until by late March their swaying ranks, swelled by the blanched faces of later narcissi, seem to form a massed band to provide a secret accompaniment to the overture of bird song. In sheltered nooks, especially along the banks of burns, primroses next raise their wan saffron moons to the misty sun, though I've never found one before 19 March, and they sag low under the weighty visits of the first humble bees. Now the tall delicate wood anemones, with their jagged leaves and floppy white and lilac tinged flowers, stage a race with the glossy yellow seven-petalled stars of celandines for pride of place on the woodland floor. In early April their ranks are invaded by the creeping rhizomes of wood sorrel whose white petals, leaning against clover-like leaves, open wide when the sun is bright but droop, the most disconsolate of woodland flowers, if it rains.

Moss, lying over rocks and stumps, is sending up new heads, and in the clearings and front meadow the first grasses, in sheltered dips or behind rocks, are greening the earth. Violets appear almost apologetically, as if modestly selecting careful places. The dark rosettes of bluebell leaves begin to starfish the ground, and by mid-April the swellings at the heads of the thick stems begin to break open. Soon the woods are carpeted in a moving mosaic of whites, yellows, greens, violets and blues. All the flowers bask in the moments of springtime sun before the trees grow their leafy canopies and cast a dense shade. Slowly the bluebells take over so the clearings between the trunks, the front meadow and the lower slopes of the hills are clouded in a blue so deep and rich it appears to herald the colour of the summer skies to come.

By mid-April too, far ahead of the other trees, the wild cherries, especially when the bullfinch tribes have been reduced by harsh winters, are covered with white blossoms. Now the well-fuzzed larches are growing crimson female flowers. Slowly the other trees transform the scene and most crowns carry a suspicion of green. The rowans are usually the next, sending out tight inward-folding spears of silky silvery green plumes and from a few yards away they look like swords pointing to heaven. Close rivals are the hazels which dangle tiny sheaves from the ends of myriad twigs, like hankies being waved to a Queen. In late April long slender buds on the zig-zag twigs of the beeches burst open and pale green leaves unfold, tiny fans fringed with gossamer. On a breezy day thousands of their light brown covering scales fall like gentle rain.

Among my sweet chestnuts caged from the winter deer, a few buds sprout small bunches of leaves while others stay dormant. The birches leaf usually between hazel and beech, throwing out folded fans of a darker jade but their small leaves grow fast, their canopy usually complete before the others. On shore the alders hasten their spiky whitish green flowers out as soon as their leaves have formed and before the end of the month they buzz with sound as humble bees compete for the tiny drops of nectar at their bases. The mighty oaks and older ashes are the last to complete the springtime scene change, their buds thickening into lumpy knobs all through April and well into May. The ashes push out the lightest green leaves of all, in small sprays, then the oaks follow with oddly weak yellow leaves, which become light khaki before turning to their final dark green.

All the while, since mid-February, the mistle thrush has dominated the morning bird song, fluting loudly from 4 a.m. to 10 a.m. (I could have shot him at times!) then seeing out the day with a long burst before dusk. One year, when no rival appeared, a single 'stormcock' took over both woods, starting in the tallest silver fir in the west wood, flying to the north-east edge of the east wood for another burst, and finally to the topmost spray of the huge spruce. Lit up by the rising or dying sun, he looked like a golden weather vane. After he had nested in an ivy-clad oak his territorial zest died a little and he was content with a brief morning piping, voice at half strength, in the lower branches of the spruce. While he preferred to sing from conifers, the song thrushes and blackbirds warbled melodiously from the flat branches of oaks and ashes.

Before all the leaves are fully grown the woodlands are bathed in brilliant green light, an inspiring stage on which animals, birds and insects conduct courtships and rear their families through the warmer days ahead. At dusk in early March the tawny owls, infrequently seen in winter when they range a wider area, become more vocal as they lay stronger claim to territories. Some nights I go out and hoot back in fair imitation and occasionally they fly nearer to investigate though they keep well out of sight. Now I have glimpsed the two great spotted woodpeckers, flashing their vivid black, white and red colours through the bare glades, to work the older hazel trunks, dead ash snags and fruit trees I have not felled mainly for their sakes. They arrive dramatically, like demon kings in a pantomime, folding their wings with all the flourish of a villain's cape.

In early April I often hear the male drumming the snags, lighter here, noisier there as he shifts about to find a really resonant spot. Having found it, he beats out a machine-gun tattoo, knowing the louder he drums the more chance he has of attracting the female. I stalk the sound and sometimes see him – traversing a tree *downwards,* hopping towards the earth and staying his fall by sticking his spiky tail feathers into the bark crevices, grabbing hold with his two-up two-down zygodactyl claws, and probing again. Once, when I saw him flicking bark off *sideways,* I thought he might be gaining relief from the usual head-aching method of banging in directly.

Each year I try unobtrusively to monitor the breeding success of the woodland birds, and in the first six springs I roughly documented these totals of young reaching fledging stage: sparrow-hawk – 2 nests, 5 young; kestrel – 2 nests, 6 young; hooded crow – 2 nests, 5 young; woodpigeon – 1 nest, 2 young; mistle thrush – 3 nests, 6 young; song thrush – 3 young but found no nest; blackbird – 5 young but only found one nest; robin – 7 nests, 31 young; chaffinch – 4 nests (hard to locate as invisibly made on branch forks, so there must have been many more), 33 young; great tit – 7 nests (all but one in my nest boxes), 29 young; blue tit – 2 nests, 8 young; coal tit – 9 young but found no nests; willow warbler – 7 ground nests, 27 young; wren – 11 nests, 33 young; woodcock – 1 nest found, 7 young seen; hedge sparrow – 3 nests, 3 young seen; great spotted woodpecker – no nest found, 3 young seen; goldcrest – only one nest found but many young seen (hard to count Britain's smallest bird when high in canopy).

Of course, I didn't find all the nests, and the woods themselves totalled a mere seven acres. Naturally, less than a quarter of the young of these smaller birds survived to breeding age, falling victim to fox, owl, hawk, and a few to my own wildcats, as well as dying in hard winters. Some nests failed when parents deserted, such as a great tit's nest being flooded by rainwater coursing down a mossy trunk, or when a brooding female lost the mate which was feeding her and had to find her own food so that the eggs grew cold. Such a bird seeks self-survival first, thus she can pass on her genes, either by a second breeding with a new male or waiting until the following year. The birds which used the bird table survived winters best while the little wrens suffered most. The wren's high breeding rate is essential for the species to survive. After the sixth and coldest winter I found no wren nests at all in the spring, and saw only three wrens in the whole year.

Next year I found only one nest in which the five naked young were dead. It takes wrens more than just one spring to recover from a severe winter.

I was puzzled by a blackbird nest deep in the rhododendron by the path, for although I saw no birds near it when I reached in I could feel many eggs. On 2 May, sure the nest was deserted, I removed one. It was a hazel nut! A woodmouse was using the nest as a larder and all the nuts had near quarter-inch holes chewed in their bottoms. Smart fellow, the woodmouse – for while he's a favourite tidbit of the fox, or slower badger, he can escape both by climbing high, as agile as a tiny squirrel, an ability that helps make him our commonest mammal.

One fine spring afternoon I was boating up the loch with Moobli and supplies when I suddenly became acutely conscious of my surroundings, as if seeing them for the first time. As I neared Wildernesse, the engine noise seemed an affliction. I turned it off, slid the long oars into their locks and rowed the rest of the way. We seemed enveloped in ancient silence. The sun was setting behind the mountains, the sky had become a vast amethyst without a cloud to be seen, and the tops of the hills and tree tops were tinged with coppery light. Such beauty, appearing suddenly after days of mist and rain, makes the lone human take stock, to know yet again it is fragile, ephemeral, fleeting, and because it will soon be gone, to be more intensely cherished – like a sweetheart in wartime.

That night dusk did not give way to darkness and as supper cooked I went outside. An almost full moon was dazzling a silvery gold path across the shimmering waters. I looked up to see what resembled a giant bat fluttering above the trees and heard faint squeaky cries of '*kissick kissick*' . Then I saw the long beak, pointing downwards – a woodcock.

There are few more beautiful sights in British nature than a woodcock performing its aerial ballet, its springtime 'roding' flight, and here was a male beating the bounds of his woodland parish. On shallow beating wings he floated eerily along, over the cottage, round the westerly pine wood, along the loch shore and back round the larch wood again. Sometimes he went into a mad spiralling dive, as if he'd been shot, only to recover equilibrium easily and carry on again.

The woodcock advertises the bounds of his territory in this way and also uses the flight in courtship to help arouse the female – before spiralling down again to land near her and offer a plaintive croaky squeak, hoping to have his way! This flight is different from the one it uses when flushed from a wood. Relying on the incredible camouflage of the rich browns, russets, yellows, greens and greys of its plumage to match the fallen leaves and debris on the forest floor, the woodcock sits tight until the last moment. Only when it knows it has been seen, or is about to be trodden underfoot, does it fly. Then it thumps down and is away like a bronze arrow, sometimes silently, sometimes its wings making a sound like ripping paper, dodging fast and low round dense bushes and trees as if knowing that way it can't be shot (or photographed!). For this reason, apart from its dark delicate flesh, it has long been a prized 'sporting' game bird. Only twice have I seen one before it saw me in my woods, the first time only because Moobli was scenting the air in a way that told me something was ahead. The bird had its beak deep into the litter, twisting from side to side as if feeling for food, and as I watched it hauled out a huge worm in a slight tug of war, able to hold it easily in the prehensile upper mandible. Naturally, I had only the standard lens on the camera.

It is a clever bird. It has learnt that by stamping on the ground, imitating the patter of rain, it can bring worms near to the surface. It is clever enough to keep ahead of frozen ground in winter and migrates south of its more northerly breeding haunts. Often there seem to be more woodcock about in early spring, not because the leafless trees make them easier to see, but simply because there *are* more, populations being swelled by winter migrants from the icier regions of northern Europe.

It is a cussedly difficult bird to study in the wild, and seems to have been put together when the Creator was feeling humorous.

Its skull appears to have been put on backwards, and its great black pop eyes are set behind its ears, so far back they have binocular vision all round the bird at any time. Even when digging in the ground with its absurdly long beak, it can see danger from above or behind.

Predators don't affect its numbers much. A prowling fox or member of the crow family may chance upon the four chestnut and grey blotched eggs, which are laid in a mere scrape among the dead leaves, but no woodland bird of prey is fast enough to catch a woodcock that sees it coming. The chicks are more vulnerable but they are superbly camouflaged too. Only once have I found a woodcock family. It was on 12 July, when I heard piercing common gull cries coming from above the woods at a favourite deer crossing on the burn and wondered if they had found a young fox or wildcat. I hurried up there with Moobli and at first found nothing. Then, as we moved through the bracken where they had taken refuge from the gulls, three young woodcocks flew up with wildly beating immature wings and drooped tails, to drop down out of sight again in a few yards.

Although I had glimpses of the lonely roe doe through the fifth year, and saw the new buck twice, it seemed they had not met and the buck was keeping to the woods that began east of the burn. But on 5 June the following year I went out of the cottage with Moobli, and he instantly looked towards the east wood and stiffened, his tail up like a thick hairy flag. Motioning him back, I crept forward and there in a clearing beyond the spruce were the roe doe and the large new buck, his antlers fully grown and clear of velvet. They were browsing so intently on the low leaves of bushes that I sneaked back for my camera. When I returned they had gone down to the shore and were walking westward through high bracken, making a clear photo impossible. I didn't stalk them in case I scared them away.

Not until early July did we see them again. We were returning from a hill trek, had just left the long wood a mile to the west, the southwesterly breeze blowing from behind us, and were stepping through the thick crowning grass of tussocks on a down slope when I saw a reddish form ahead. Sure it was a big fox, I dropped to my knees, fixed on the telephoto lens and tried, with the wind wrong, what seemed a hopeless stalk.

Suddenly the animal looked up – it was the roe doe. She got our scent, strong after the perspiring trek, and almost did a somersault to get down into the bracken again and shoot away. As she ran the buck appeared as if from nowhere and went after her. At about 200 yards they paused to look back on an open slope and I got a few pictures. Off they went again, vanished through some trees across a burn, and we dashed madly over the tussocks to get nearer, but they were still too far away. With bounding ease, like giant hares in red summer coats, they traversed the steep slopes and disappeared. I was so surprised to find them out in the open bracken in daylight, and cursed myself for not keeping a more watchful eye ahead. By cutting higher earlier, we could have got close.

It seemed the roes now had a special territory in the long wood, but one morning in late July I found one of two small birch saplings in the west wood had been frayed bare of leaves. Some bark had also been stripped off and was hanging in shreds from broken twigs – a roebuck's work. Below the sapling were the tell-tale droppings, darker, more cylindrical and smaller than those of red deer, and dented at one end. I kept sporadic watch but was not rewarded until just before dusk three days later, when I heard a high-pitched bleating noise.

I sneaked out with the camera, into a light damp mist, and stalked towards the west wood using rhododendron and hazel bushes as cover – to witness a wonderfully intimate scene.

In the small clearing containing the two little birches the big roebuck was 'driving' his doe in front of him. Dancing along in a figure of eight round the two birches and a small fir tree, they seemed oblivious to danger. The doe was doing a fast sort of creep-walk as if torn between the desire to flee and the desire to mate, and his black muzzle was merely inches from her rear, the white powder puff of her alarm patch bobbing tantalisingly before him. He followed her intently, not actually trying to mount but as if to enjoy her scent to the maximum first. His nine-inch antlers were pressed back against his neck and his eyes were half closed in what seemed a leer of passion. He was making odd raspy grunts. This went on for two minutes or so while I bemoaned the lack of light and the wet mist, knowing it futile to try for a photo, but still hoping to see the actual mating.

I had no such luck. A heavy shower hit the wood, heavy drops spattered down from the high leaves, and this seemed to alarm the doe, which stopped short, then ran off through the trees and vanished. The buck looked at her bobbing white flash, looked at the untouched birch tree and then, as if in irritation, started to fray it. He bent his head down, thrust his antlers between the small branches and with fast up and down heaves frayed off the bark and many of the leaves and upper twigs. I was glad then, not only because I had caged my small new trees but because I had left plenty of such 'weed' trees for the deer to ruin.

Bucks fray like this first in May to rub off the velvet from maturing antlers, then more often before the rut to smear scent from their forehead glands on the white stems they have thus debarked. They do this to mark their territories – usually decided by the favourite area of the doe – which can be as small as seventeen to twenty-four acres where feed is plentiful. In my area, where woods were thinly strung out along a loch, this pair ranged over three miles.

When the buck got his tines stuck in a branch, he twisted his head right round, then danced in a circle with his back legs, as if trying to snap it right off!

Suddenly there was a movement. Moobli, whom I had left in the cottage, must have nosed open the half-closed door and had come to see what I was doing. The buck may have heard the scrape of his claws on a rock for he swivelled his great mussel-shaped ears round nervously, then bounded upwards after the doe. I couldn't blame Moobli as I had not actually ordered him to stay put. We sneaked down to the clearing. The roes had made a real 'roe ring', a worn figure-of-eight track round the birches and fir. Both birches had been hopelessly mashed by the buck's antlers, some branches twisted round and round, as if a human had tried to screw them off.

Beautiful though they look, the dainty roebucks can do great damage in young plantations, for they do this scent-anointing fraying on young trees all round their territory and along paths to their favourite feed areas. If another buck moves in they also fray in threat display before fighting off the intruder. When a mature buck is shot or dies, young bucks moving in to compete for the doe fray worse than the resident buck would have done.

In browsing, roes eat the growing lead shoots, so crippling young trees. They also eat seedlings, so preventing natural regeneration. Being small, they are harder to keep out of enclosed forests than red deer. Once I saw a roe doe crawl through a ten-inch gap below a fence.

In Scotland roebucks are protected from 21 October to 30 April inclusive, and the does from 1 March to 20 October inclusive. But both can be shot if causing extensive damage on enclosed lands. However, the best keepers value these delicate forest nymphs which, with red deer, are the only two of our seven wild deer species which are truly indigenous to Britain. They would never shoot roe in the rutting season, or does in the breeding, birth and weaning season, legal in enclosed woodlands or not.

We stole upwards through the trees beside the big rock escarpment and hadn't gone far when Moobli's great ears pricked forward. There, in a small gap, was the roebuck again, apparently *dancing!* He was up on his hind legs, balancing like a ballet star, front hooves dangling daintily, neck craning to eat the lead shoot of one of my young oaks. I raised the camera. *Click.* The buck heard it before he got the shoot in his mouth and bounded away. Another fuzzy photo in the damp Highland glimmer, but the sightings alone had been superb.

I caught occasional glimpses of the roes after that, usually from the boat on supply trips, and the following May I saw them with two well-grown yearlings, so they had at least mated successfully. One thing puzzled me – as roebucks are so fierce towards each other in the rut in late July and August, what happens to any youngsters born in May or early June? I have not been lucky enough to find out. A keeper friend, Peter Madden, whose fascinating job consists of keeping some 50,000 acres of Forestry Commission woodlands in the region, eventually came up with the answer.

'One July morning I saw a buck chasing his doe in a figure eight round some small rowans, and she had twin fawns about two months old with her,' Peter told me. 'Evidently the buck had no paternal feelings towards them in the rut, or he may have been a different father. But he went for them, to drive them out of the way. Both youngsters dived into the nearest bracken, and he followed as if trying to scent and listen for where they were.

'In the bracken the fawns split up and shot out of it in two different directions and ran off. For a while the buck just stood there, shaking his head. He didn't seem to have a clue where they had gone.'

Evidently the fawns have to learn to keep out of a buck's way during the intensive period of the rut, and seek their mother's milk when they can.

Peter told me that because roebucks do far more damage to young trees if two or more are competing for a doe, it was bad forestry policy to leave too many bucks. Once he saw a mature buck chase a yearling buck off his territory, and attack it fiercely with his antlers. When Peter shot the yearling he found it had two bleeding puncture marks in its right rear haunch as if made by a drill.

The roe is unique among British deer in that the doe enjoys delayed implantation of the blastocyst: the fertilised egg doesn't start developing in the womb until December, giving her the ten months' gestation of a much larger animal. If a doe doesn't mate in the usual rut she can do so during a 'false rut' in the autumn. Yet she can still drop her fawns, usually twins, in late May and June, the fine easy season. Why roe deer have developed this capacity seems just another of nature's many mysteries.

Chapter 15

The Red Squirrels

If I ever had to name the real spirits of the woodlands I suppose pride of place would go to the red squirrels. The first creature I saw on my first boat trip to Wildernesse one gloomy February morning was a red squirrel dashing over a ruined wall – which helped clinch the place in my mind as a good area to live. It was mainly to help them that I planted the sweet chestnut trees.

Through the years and all seasons, for contrary to popular belief they don't hibernate in winter and can't last more than a few days without food, I have studied them – a hard task in the wild. The moment I got near one, off it would dash, pause briefly on a larch or oak for some scolding '*kok kok*'s like someone knocking on a door, then foil my attempts at photography by climbing up the *far* side of the tree, so all I got was a brief glimpse of a whisker, foot or end of its tail as it deliberately kept the trunk between it and myself.

Time and again I passed under trees, heard a scuttering, and looked up to find a squirrel had heard or seen me coming and had flattened itself, tail and all, along a limb to escape detection. For a year all attempts at a good picture failed. Then on 9 September I saw a red squirrel zipping through hazel bushes for nuts, then leap on to the trunk of a large ash. With telephoto lens, I stalked it like a deer but its movements were so fast I kept failing to get it in focus before it was away again. Finally it stopped for a few seconds, high up but in clear view just as a beam of sun shone through the clouds, and I got my first clear shot.

Only then did I realise there were two squirrels: the one I had seen a week before had a reddy-orange body and bushy yellowish tail but this one had a much darker pelage with black hairs in its tail. Sneaking away, I hid behind a hazel bush but kept a keen watch. Slowly it worked its way back to the west wood and through the fieldglass I traced it to a large drey, high up in a dark twin-forked crotch of a tall silver fir. Like all red squirrel dreys, it was tight to the trunk and had no obvious entrance, the foliage closing behind as the animals went in or out.

On 1 October the darker one I thought to be a male showed me how a red squirrel gathers nuts, when he gave the hazels in the front pasture a real going over. The phrase 'forest sprite' seemed inapt because, for his size, he had a burly, powerful body, seemed tireless and to be nearly all muscle. His big eyes bulging with intent, he grabbed a branch with one hand, pulled it down to see if there were any nuts there, then let it go. Sometimes he waved a branch, or thrust it away with an aggressive push as if to help him on his fidgety way. As he jumped from one to another I saw that he kicked against already rising branches to give his upward leap more impetus, just as he kicked against the down-moving branches to reach one below more quickly. He worked with blinding speed and was baffling to watch.

He dropped all the nuts to the ground, dashed down the trunk, stuffed them in his mouth and cheeks with his 'hands', then ran to stash them somewhere near the biggest ash tree on the shore. Then back he bounded, bushy tail streaming out behind like a banner, up another bush, dropped two more nuts, then down again. He seemed able to count for he shot up another bush, dropped one nut, then as if knowing he had all he could carry, he dangled momentarily from both hind feet, dropped to the ground, and carried that lot away too.

A few days later I found this hiding place, at ground level in a hollow between old hazel boles, which he seemed to have partially covered with moss. Red squirrels cache food by instinct in periods of surplus but have short memories so have to relocate them by search and scent. Many seeds and nuts are forgotten so the animals become unwitting forestry 'planters' for such seeds as hazel nuts, acorns, beech mast and those from pine cones. He must have forgotten the hazel bole cache for in early winter I found woodmice had moved in, the holes in the nuts too small to have been made by squirrels, which would have removed the nuts anyway.

I soon found they didn't confine caches to ground level but made little hoards in tree holes, mossy forks – almost anywhere that took their fancy, even single nuts stuck in bark crevices of the giant spruce! On several autumn mornings I was woken by loud thumps in the rear of the cottage roof. Not until I dashed out and saw a red squirrel leap from the far guttering and run up an ash did I solve the mystery: they had been caching nuts in a gap between the wooden roof sills.

In the first months I made several bird nest boxes, the pride of which was a large 'owl' box, more than two feet long, which I laboriously installed at fifteen feet below a sheltering branch of a large oak tree. Although at dusk in spring I could hoot the tawnies to its vicinity – hoping they would discover it 'naturally' – no tawny ever used it. On 6 February of the second spring, I saw a small twig in the entrance of the box. I climbed up with a ladder and found the box half filled with budded hazel and birch twigs, yet I never saw a squirrel near the box.

In mid-March of the third spring, when a tawny owl had been hooting near it, I checked the box again. Now it was half-filled with dry moss and thick-budded oak twigs and as I carefully put my hand in I heard a harsh chattering sound. Only ten feet above my head the male squirrel, in his darker winter coat, was scolding me with his explosive chucks, flicking his tail agitatedly, thumping his little furry clawed feet down on the branch – even, I fancied, baring his little buck teeth. I knew then it was indeed the squirrels who were using the box, storing the budded twigs for winter food. It also seemed from the moss that they were using it as a secondary drey when foraging in the east wood.

Moobli found squirrel scent as fascinating as that of fox, deer trails or wildcats, and often I only knew the animals were nearby when he suddenly put his nose down in a clearing, followed it briefly to a trunk, then looked up into the tree with excited expectation over his face. On 14 July in the third summer I followed one such gaze and saw the male squirrel, looking a trifle thin, up a beech tree. He ran along an upward snaking branch, appeared to slip a little on the smooth grey-green bark, until he was lost amid the dense upper leaves.

On 16 August we were heading into the west wood to collect some second-crop golden chanterelle fungi for hors d'oeuvres, when Moobli alerted me the same way. This time the female squirrel, fat as a butter bean, was up an oak tree and as I looked she dropped what she was holding and leaped for a pine that seemed impossibly far away. Her fluffy tail flattened out, her four feet were extended at right angles so that the loose skin between them and her body made a slight sail (it was more a glide than a leap) and she disappeared into the thick conifer needles with no trouble at all. Just then what she had let go fell almost at my feet. It was a chewed piece of golden chanterelle.

When I reached my little patch I found every one of the fluted trumpet toadstools gone! On a stump nearby stood the remains of two more, with deep incisor marks in them. Next day I actually saw the female squirrel pick one. She plucked it, heaving up with both hands like an animated cartoon squirrel, chewed it like a corn cob, then also left it upended on a mossy stump. So now I had to compete with the squirrels, as well as deer and badgers. Well, one wilderness law seems to be Do Unto Others Before They Do Unto You (or Before You Get Done), so I had no complaints. When I found a huge late crop of chanterelle on the steep banks of the burn gorge some 300 yards above the cottage two weeks later, I uprooted some carefully, shook out their spores then replanted them in the west wood. The squirrels luckily spurned this human scent-tainted crop, though we had the biggest ever yields there the next year.

In December of the fourth year I was perplexed by little pits dug in the snow in the west wood. While we often found the big twisted whorled holes made by badgers digging for grubs and worms, these were much smaller and too late in the year to be made by young badgers. Not until a heavy snowfall in early April did I solve the mystery. I woke to find overnight blizzards had left four inches of snow over the ground and the daffodils were looking strangled, up to their necks in it.

In the glade under the spruce where the snow had been driven to form an even white carpet, were tracks that could only have been made by a squirrel. It had come down the scaly spruce trunk, landed in the snow mainly on its back legs and then hopped with its back feet apart in a V like a little kangaroo, with the front paw marks neatly in between. The tracks straightened after a few yards and went to below a hazel bush. There were the odd little diggings again and I found a few hazel nuts that had clearly been dug out of the litter of autumn leaves. The nuts had been split in half by the squirrel. About forty-five feet up the dark spruce a small new drey had been made.

It was now obvious, and interesting, that the red squirrels were working both woods through the year, living by choice where the most seasonal food existed. Their main diet was conifer foods – Scots pine cones, seeds, buds, shoots, young needles; a few shoots, seeds and buds of spruce, larch, fir and other pines; some buds, shoots and green leaves of oak, beech and birch. In late summer and autumn they feasted on berries, beech mast, hazel nuts and acorns, found several fungi a delicacy including the yellowish tricholomas which were common and which they again left upturned on stumps. They also ate some larvae, pupae and insects. In winter, apart from their own caches, they dug for half-buried nuts, acorns and beech mast, and when the first buds came they took the tips of several broadleaf trees.

It was in the autumn of the fifth year that tragedy befell the little squirrels. At dawn on 27 October I was woken by a scratching and thought it was just some chaffinches scrapping on the bird table. I looked out into the dewy dampness and there at a piece of bread was the male squirrel, but looking very bedraggled. His zippy movements were slower than usual. I slid silently over the floor for my camera and took several photos of him. I noticed his eyes were half closed as if inflamed. He looked really old, his tail sparsely haired and his teeth long and brown, not well at all. Then he dropped to the ground and went hoppity away.

In the afternoon Moobli and I were walking through the east wood when he picked up a scent. I looked up to see the squirrel in the largest oak tree. He had seen us and was leaping from branch to branch to try and reach the larches. Suddenly he hit the edge of a broad branch, scrabbled at it in vain, fell to the ground with a thump, bounced up, but seemed unable to get under way again. Moobli was almost on him. I yelled, 'Na? Come here!' The squirrel turned on the dog, terrified but ready to fight if he had to. Moobli had no wish to hurt him and kept back at my command. Yet instead of running away the squirrel kept stumbling in a half circle. I reached him and got him by the tail and put him into the spacious cage I had once used for my wildcats.

To my horror he just rolled on his back, stretched out all his legs and claws, opened his mouth wide, gave a slight shiver – and was dead. It was an astonishing and sad occurrence. When I examined him I could see he was not only very old, for his teeth were long and loose, but he probably had a disease too for there was a slight yellow discharge from his nose. He must have been ill or he would never have come to the bird table for bread, nor would he have missed his footing on what for a squirrel was an easy jump. Doubtless shock and a heart attack from the fall and our close appearance had been the final factors. I weighed him: just over half a pound. Then I measured him: head and body 8¼ inches, tail 8½ inches. But as I held his warm little body, seeing his dark brown ear tufts and long red-brown whiskers, I felt sad his life had ended this way.

We did not see the female squirrel all that winter or the following spring either. Red squirrels can suffer from sudden declines, usually after a failure in their main food crops, quite independently of the presence of grey squirrels (which did not exist in my area anyway), so it seemed they had both had a bad year and she too had succumbed to a virus.

I wondered if I should catch some squirrels from an abundant area and put them in the woods, but decided if there was a virus about it would be better to wait a year or two. As it turned out I had no need, for in spring something remarkable happened.

On 5 May Moobli and I were out on the steep open hills half a mile to the west, trying to set a live trap for my old tom wildcat. A movement caught my eye and there, only thirty feet above us, what looked like a small pine marten was working its way between the rocks eastwards. We kept still until it showed its whole body momentarily and I could see by its fluffy tail and ear tufts it was a dark-coloured red squirrel. It was sneaking with fast flashy little darts along the rock ridges, slipping under heather, flattening out over rocks and into crevices as it kept looking nervously about. I was sure it was heading for our west wood.

Was it deliberately going to fill the gap left by the male that died seven months before? If so, how did it know? Was this how squirrels filled new territories? Could it see for up to a mile, for how would it know the wood was even there? It went against rational behaviour for a red squirrel to expose itself to danger over a mile of virtually open ground unless it was in search of new territory and the conifer trees it needed. There had been an east wind for several days, so perhaps it had actually scented the pines. Of course, carrying the trap with me, I had no camera, so couldn't record the event.

Deliberately I did not keep searching the woods to see if the new red squirrel had set up residence, but on 27 November, the second of two gorgeous sunny days, I saw it. Moobli and I were just setting off on a hill trek when I spotted the squirrel, distinctive for its same dark colour, gathering nuts under the hazel trees near the giant spruce. I even took two photos of it.

All red squirrels grow darker winter coats, with blackish hairs in the ear tufts and tail. Our new resident was so dark that were we living in an industrial area I might have believed it to be a melanistic form. But there was another explanation. Due to the destruction of Scotland's great natural forests after the Middle Ages – to eliminate wolves, drive out rebels, provide charcoal, fuel for smelting ores, lumber for building and large ships' timbers, and to make pasture land – the red squirrel, hindered also by periodic disease, was almost extinct by the early 1800s. The more enlightened estate owners then began importing large numbers from England and a few from the continent, among which was an extremely dark brown form of the species. Where they met these freely interbred with the few indigenous Highland red squirrels still left, so the colour variation today is considerable in some areas. I was sure our new animal was a descendant of those first continental varieties.

During the sixth and harshest winter at Wildernesse I frequently saw the squirrel in the glades and, to help it, I unobtrusively threw nuts over the carpet of leaves and litter. The old drey in the spruce was blown out by the gales but the squirrel was clever enough to use most of the twigs again to make a new drey in a larch which stood in the lee of the giant spruce tree, free from the prevailing winds. By early spring the drey was twice the size and like a rugby ball. I reckoned the squirrel was a youngster, and tough, for many of the empty nuts I found had been split in half without a hole being gnawed in them first.

I was writing at my desk on 2 June when something moved outside the window. A large bright red squirrel with a long flowing cream-tipped tail came loping across my view, paused at the bird table, then went on towards the rhododendron bush by the path. It happened so fast I had no time to get the camera. It didn't look like our resident squirrel, being much lighter in colour and also bigger. That afternoon I was lucky to get a closer look at it.

I was walking down the path to the boat with Moobli when he suddenly gave a slight 'whuff', then put his nose down and followed a scent to the same rhododendron bush. Thinking it might be my tamed young wildcat, Liane, returning from one of her springtime expeditions, I parted the thick leaves. Instantly an animal dashed out of the far end, over the pasture, then leaped for the lower branches of the largest cherry tree. I glimpsed the bright red-brown body and heard its sharp '*zut zut*'s as it climbed to the top.

I raced for my camera. As it flirted its tail jerkily and scolded us from the upper twigs, perhaps not daring to try the long leap to the nearest larch of the west wood in case it fell to the ground, I managed to take nineteen pictures of it. I thought then that it just might be our resident squirrel, for it acted as if it knew us, performing acrobatics in full view as if deliberately posing for the sort of photos I had been wanting to take for six years.

Two weeks later the little saga of our red squirrels had an even happier ending. I was collecting some of the first crop of golden chanterelle in the west wood when I heard a scratching sound above. There, up in the canopy of the pines, were *two* squirrels. The bright red-brown one with the blond tail was leaping from branch to branch, quite slowly and certainly not in fear, while behind it, as if in laconically amorous pursuit, went the dark brown squirrel which had been our resident for a year. I was delighted the little woodlands once again contained a pair of these engaging sprites.

There are no more lovely places in the countryside than woodlands in the glorious upsurge of spring. It is not merely for the inspirational beauty and spiritual solace they offer, nor because they are wonderful theatres in which many forms of wildlife are the strolling players, that we must work harder to enhance the forests we have left and plant new ones. They are vitally *useful* to us and all forms of life too.

Trees are, with the phyto plankton of the oceans, the main lungs of our planet, absorbing carbon dioxide from the air in sunlight and releasing oxygen. Woods are natural windbreaks, protecting crops, homes and animals in their lee and sheltering many kinds of wildlife in winter. Their roots prevent landslides on hills, soil erosion from wind and rain, and also filter the waters that drain into our rivers, lakes and reservoirs. Evaporation from woodland soil with a deep leaf mould layer is less than a quarter of that from open land. Forests help make the air moister, warmer at night and cooler during the summer day, producing a more equable climate generally, and they are ideal nurseries for many young birds and animals. When I looked at the little broad-leaf trees I had planted and caged I realised it was a mere drop in the ocean of what man needs to do to reverse the tide of destruction he has, and still is, meting out to the health- and life-giving wooded wildernesses of our earth. And we must do far more than follow previous short term policies of the main forestry agencies who have mainly planted close-packed artificial stands of quick-growing 'economic' conifers for pulp, paper or pit props.

Forests planted with varied hardwood broadleaf trees such as oak, chestnut, beech and walnut among the conifers will both enhance our wildlife *and,* if well managed, provide us with all our timber needs and fuel too – an increasingly important consideration in itself, assuming man doesn't blow himself up before the oil runs out. We have enough coal in Britain alone to last three hundred years, and we can make oil from coal if we still need it then, and of course, coal itself comes from trees.

In the afternoon of 9 May Moobli and I went to look round the two little woods. The sun was shining from a gloriously cloudless sky and the first bluebells were springing from the pasture. Two merganser ducks and a drake were idling along the brimming surface of the loch and from the shore the soft cries of the newly returned sandpipers came through the trees. In a clearing in the west wood I saw the first green hairstreak butterfly land on a bramble leaf, closing its little wings to reveal an iridescent patch of emerald green beneath, then a speckled wood flopping precariously between the trunks.

Out over the grassy pasture I had created between the woods by yearly cutting down the smothering bracken, the first green veined whites were wisping like tiny flecks of light. All round us the springtime woods were filled with music – the see-saw songs of tits, chipping trills of finches, shrill pipings of dunnocks and somewhere far above I could hear the ringing cry of one of the buzzards. Now the day's first bats had come to work the air in the glades, their wings glimmering pinky-brown as the sun's rays shone through them.

As we emerged from the wood again the two cherry trees were a mass of pinky-white blossoms and seemed to be humming. I counted thirty-seven ponderous humble bees bumbling over them, then saw a peacock butterfly alight to probe for nectar, its wings kept proudly open showing colours more brilliant than any rainbow. After the dark cold winter it was now in the sun on beautiful blooms.

I felt heady with the beauty, the wonder of the flowers, and as I gazed entranced an evocative phrase by the Cretan writer Nikos Kazantzakis came to my mind:

> *I spoke to the almond tree*
> *Sister, speak to me of God*
> *And the almond tree blossomed.*

PART FIVE

Chapter 16

The Mountains in Spring

In the early years 'The Hill' (as Highland Scots call their mountains in the plural) was to me mainly a challenge, a way of keeping fit. Each morning on the days I had set aside for trekking, I would go out, assess the weather, look at the 500-foot ridges towering steeply above the cottage knowing they had to be beaten first but were merely the prelude to more ridges, higher slopes and beyond them more crests to peaks of nearly 3,000 feet. And my heart would skip a beat as I knew what lay ahead.

Within an hour I would be miles into a wilder landscape, forcing through deep tussocks where flowing grassy crowns disguised the treacherous gullies between them, skirting spongy bogs with floating beds of sphagnum mosses, floundering over black peat hags, negotiating steep rocky scree where a false step meant at least a bloody scraping fall, climbing rocky ravines carved over thousands of years by the burns, and pushing through high bracken that cut hands as painlessly as an anaesthetised surgeon's knife if I grabbed it to arrest a slight fall.

I recall one late March trek in the first year. I had climbed to 1,500 feet, then down again and along the river of a steep glen for three miles, up to a 2,000-foot peak, back along the high saddle for a mile, then back down again to the river valley. I had seen nothing but the first primroses on a burn bank, a few hinds, a pair of ravens flying together, a few molehills at the unusual height of 1,500 feet, and was then on my way back up out of the glen. What a dreary slog it had become – one foot in front of the other, force the creaking painful knee straight again, then another step up, time after interminable time, zig-zagging yet again up the steep jagged ground. Miles of mostly staring down, every tussock, rock and slab of embedded granite beating its unseasonable heat back at me, my pack of camera, lenses, rainsuit, heavy on my back, and my only consolation seemed to be if I fell now, going up, at least it would only be a matter of inches to the ground.

Finally, for the first time in my life when I did not actually choose to, I had to sit and rest. I was forty-six and wondered if after a fast-travelling life as journalist in many of the big cities of the Western world, after the hard Canadian wilderness years, I had become too old physically for what I was attempting. Yet I was reading about 'old' men of sixty who could cover three Munros (3,000-foot peaks) in a day!

Still, all this and more had to be endured if I was to see wildlife at its finest in the mountains – the bachelor herds of red deer stags avoiding flies on the high plateaus; an eagle with wings tucked in dropping out of the sky a mile above, lit by the sun into a ball of gold, shooting upwards again after a brief opening of its wings, then repeating the deep undulating dives over two or three miles in a spectacular territorial display; a tawny wildcat slipping between boulders of a high rockfall; a distant fox beating long grasses with forepaws to shake out grasshoppers; a small pack of ptarmigan whirring over a ridge, snorting like little pigs at this rare intrusion into their lonely alpine world; or if I was to emerge finally upon the highest peak and survey far below a world of timelessness and mystery, where mountain followed mountain, and glen succeeded glen in a shimmering vista that seemed to have remained unchanged since time began, inspiring thoughts that would never have come at my desk.

It was a long time before I realised my mistake. Firstly, I was travelling too fast, trying to prove to myself I was not yet 'done' on the Hill; and secondly, that the mighty 'Munro baggers' did not descend to *sea level* between each peak, but chose ranges where the drops were seldom more than 500 feet. In the end I learned that the way to trek steep open hills was to be easy in mind, rest my heart, go slowly with shorter steps, and never top a crest or pass round a knoll without pausing for a slow look ahead first, to move like a hunting wildcat. I learned that I should not walk to likely places and then start stalking wildlife, but from the second I left the cottage to be cautious moving all the way. One could never say the Hill was empty, although it appeared to be, because one never knew what animals or birds were hidden in the clefts of such steep undulating country.

Above all, I came to realise I should not look on the Hill as a stern taskmaster, so I imagined myself trekking beside a real Zorba-like Highlander who gave me a phrase which I tried to incorporate into my trekking soul.

'Don't fight the hills, laddie! Be part of them. Make no effort. *Allow* the hills to take you up!'

Trekking for wildlife involves stalking, and when one is travelling over mud, sand, loose shale, wet grasses (which remain trodden down longer than dry) or winter snow, tracking too. From the years in Canada and on the sea island where I had first lived in the Highlands, I knew the basics of stalking – to work when you could with the sun or greater light behind you as it helped dazzle the creatures ahead, against the wind as it took away your scent, and that this too travels further in warm or humid air than in cold or dry. Such basics are not enough, however, for every hill is different and during the first six years of exploring some 300 square miles I learned slowly, often humiliatingly, the vagaries of the wind.

In a strong north-westerly it might seem right to trek against it up a three-mile river valley that runs north-west from the loch shore. But that wind would come through a gap in the hills four miles to the west, strike the steep mountains on the south side of the loch (which run southwest to north-east) and be funnelled by them in a north-easterly direction almost as strongly as if a south-westerly were blowing. Once in the river valley the mountains on the east side would channel the wind back up to the north – in almost precisely the opposite direction from the way the clouds were moving above. Not until you had covered two miles of the valley, when there would be a gap in the hills to the north-west, could the true nor-westerly assert itself and give you a stalking wind. It is impossible to stalk the first part of that valley if the wind is anything but north to north-easterly. Sometimes, though it takes years to learn, you can stalk *with* the wind. If you are moving up a long narrow corrie or burn gorge and the deer you want are on the right of a knoll at the top, for instance, the gorge can provide a 'chute' for the wind and direct it to the left side of the knoll, away from the deer. But always on any stalk you need constantly to check through your fieldglass which way the wind is blowing where the animals actually *are,* either by the moving of the grasses or a small hillside birch or rowan tree.

I also learned to make full use of deer paths, for red deer always follow the route of least resistance when going upwards. In snow on the tops deer seem to know the lie of the land beneath, so it was wise to keep to their piston-deep tracks – or you would find yourself floundering in three-or four-foot drifts, stumbling over deep tussocks with slush beneath that filled your boots, or falling into channels cut into the ground by small burns. When going downhill, especially in tussock country, I learned always to keep my feet turned outwards, so that the ball of the little toe and side of the foot hit the ground first. Ankles are usually sprained or broken by twisting out to the side, or more rarely to the front; it is almost impossible to bust your ankles inwards!

Early spring, before the leaves are full on the trees, is always a good time for trekking, and one early April morning when Moobli was still a five-month-old pup, we set off for the river valley to see if the stags were dropping their antlers yet. As we tramped across the slopes of heather cut short by grazing deer, the female buzzard came beating overhead, attended by two smaller males, and gave a ringing '*kee-oo*' call. As they passed above one of the males dived down and the female rolled side-ways, momentarily grabbing his talons with hers. I have only seen buzzards actually grip talons in this way twice, and I hoped it meant she was really keen on this fellow and that I would find their occupied nest.

As we turned to cut through the woods to the shore line, I saw five hinds and two yearlings grazing just to the west of our west wood. The wind was now coming from the north-east. Two days earlier, with the wind coming from the west, the same group had been grazing in the open field east of our east wood. It seemed, when on low ground, they naturally stationed themselves leeward of woodlands, not only for shelter from the winds but because they could see in front of them while also scenting any danger that might be coming unseen through the woods. As we reached the shore two dippers flew up with odd '*zink*' cries, then with strong flight that looked as if it should be whirring audibly but was in fact silent, they landed on rocks further away, bobbing their stumpy chocolate tails like giant wrens.

After leaving the wood we had to cover half a mile of deep tussocks. The new light green *molinia* had not started to grow and the long bleached grasses of last year, known locally as 'white grass' in winter, now lay flat over the crowns so it was easier to see where I was putting my feet. As we forded the main river I saw that some 300 acres of the valley had been planted with young conifers and had been fenced off against the deer. We headed up the valley, skirting slakes of still water left by the winter torrents in which frogs had laid masses of spawn. Some of the black centres were already elongating into tadpoles.

Just before the valley's end turned west and became a large high corrie, I saw some browny-black objects which looked like rocks on the slopes. Through the fieldglass I saw they were stags – eleven of them, dark in their winter coats. Halting Moobli each time he nosed ahead, and lying on the ground occasionally which gave him a sense of caution, we stalked them. As we came near I saw only two had lost their antlers, and one had only dropped one. It gave him a lopsided look. It was a late season, the new grass not yet growing, and they looked thin and weak.

As we crawled nearer, my knees soaked and bitter cold, two of the near younger stags must have got some of our scent in the now-sideways wind for they trotted anxiously left, then right, and finally went through the others until 200 yards further on. The older stags just looked up briefly, then carried on grazing. At this time of year red deer are at their lowest ebb after poor winter feeding and don't like to run unless sure of danger. We got nearer, I took a few photos in the poor light, then stood up to relieve my aching knees. They just stared at first then seeming most reluctant, broke into a slow trot for a hundred yards before walking again.

We headed home. Just before reaching our own woods I saw one of the dippers in the loch at the edge of the burn. Every time it dipped up and down on a rock it flashed the whitish eyelids that protect its eyes under water, and flirted its wings briefly like a robin. Then it began wading deeper into the loch, pausing on one foot and turning stones over under water with the other, looking for water creatures. It was doubtless very strong for some stones were about four inches long.

To my astonishment the bird kept walking out, shrinking further below the surface, until it disappeared! I had seen a dipper walk under water by going upstream, tilting its body so the flowing water pressed it downwards, and also swim underwater with its wings and legs: this was the first time I had known a dipper to walk under completely *calm* water, clearly clinging to stones to hold itself down. Yet one look at a dipper's feet (big for its size) makes it clear such a trick is possible.

A few days later we were heading up the steep ridges when I heard a harsh '*kar kar*' coming from above the west wood: the female buzzard was sailing along hotly pursued by a hooded crow. Four times the grey-bodied crow looped above, dived down to swoosh past the huge buzzard then swung up as if on a string to do it again, trying to drive her from the area of its nest in a tall

Scots pine. At first the buzzard just soared serenely on, but evidently the last dive had aroused her anger for she suddenly folded her lower wing, twisted sideways and made a great slash with her talons as the crow shot past, a safe two yards away.

It's always surprising that bigger birds don't retaliate more to this sort of pestering, but such a commotion is more symbolic than a real attack, where the crow would take a big risk of injury or death. The mobbing bird feels righteous near its nest, whereas the victim feels less so, and thus gives way.

In our third spring, however, I actually saw a buzzard turn on one such pursuer. We were below the huge 400-foot cliffs east of home, where ravens had their nest, and I was hiding beneath a holly tree hoping to see the birds. Suddenly I saw them both high above, doubtless leaving their greeny-blue brown-blotched eggs to keep warm in a patch of sunlight. Revelling in the high winds, they plunged and dived on each other. Twice the male rolled over on his side in a double twist, even holding his flight for a few moments while upside down and giving out a queer '*boing*' call, like the twang of a giant jew's harp.

This showing-off game soon ended when the male and female buzzard appeared high above their cliffs, minding their own business, hunting from east to west. Immediately both ravens set off in pursuit, beating their wings as hard as pigeons', and soon caught up to the casually soaring pair, to 'mob' them away. At first I thought the buzzards would just sail on as usual – but not this time.

Suddenly the smaller male decided he'd had enough of this. He turned swiftly, wings narrowed, and chased both ravens, now croaking with disbelief. The ravens plummeted all over the sky but the buzzard, more manoeuvrable than the big hen, seemed determined to get his talons into one of them and matched every dodging movement. The ravens finally shot into some stunted trees near the cliff top, scuffling into the thickest twigs for cover, and the buzzard landed nearby, glaring at them. After two minutes one raven took off, flying low, then the other followed.

When the buzzard took off, both ravens turned and dived at him again. Once more he chased them into the trees! All the while his mate soared airily above as if enjoying the spectacle, and then, satisfied how well he had turned the tables, he desisted and flapped on upwards to join her in their westward flight. I was sure this meant the buzzards were nesting somewhere in the woods and the male wasn't going to stand any more nonsense on what was also his territory. I just hoped I could find the nest they were using.

I thought that would be the end of such mobbings. Later that afternoon, however, I was disturbed at my desk by a commotion of harsh croaks and '*kar*'s. I shot outside to see the male buzzard being hotly chased by no fewer than four hooded crows and five ravens. How on earth had all these corvids come together in the nesting season? This time the buzzard did not retaliate but put on a burst of speed, of which, until then, I had not known a buzzard to be capable, and soon outdistanced the mob. Since then, I have been impressed by the speed at which these birds can move if they want to. One morning Moobli and I were walking along the edge of the west wood when there was a loud splashing and then the male buzzard shot out from a small rivulet almost under our feet, where it had been taking a bath. It arrowed away, twisting through the trees like a giant woodcock despite the droplets showering from its wings, almost as manoeuvrable as the sparrowhawk.

In late April the first bracken shoots were thrusting up like double loops of green ropes on the lower hill slopes, still tawny with the look of winter for new grass had not yet begun to grow. In the spring sun some huge hairy caterpillars of the northern eggar moth, browny velvet with black rings, had emerged from hibernation and were stretching out on their food plants of ling and heather. They ate for half an hour or so and then lay prone, conspicuous to the eye at several yards, as if confident birds wouldn't eat them because of their massed spiky hairs. A few bluebells, sending an expedition party from those flowering amid the primroses in the woods, chimed silently in the breeze. Wood anemones flourished in the warmest crevices and isolated wild violets reared little pansy faces, their pale purples enhanced among small beds of chrome yellow bird's foot trefoil.

At this time the first male cuckoos, newly arrived from wintering in North Africa, were making the hills echo with soft, mournful double notes as they took up their old, or looked for new, territories, seeking to attract females. At first I felt it strange that cuckoos should come all this way, to such barren-looking hills, to find victim foster parents in whose nests they foist their eggs. Yet over the years, with heathlands becoming scarcer due to more intensive farming down south, the Highlands have become the birds' main stronghold. Here too they find their main victims, the ubiquitous meadow pipits, and also one of their favourite foods – large hairy caterpillars, like those of the eggar moths, for these birds have special gizzards that can cope with the hairs. It is an odd fact that as the sound of cuckoos in spring fills the hills, the eggar caterpillars seem to know by inherited instinct that these birds are dangerous for they become far less conspicuous in the sun. But come October, when all the cuckoos have migrated south, they can be seen in large numbers again.

I once saw how a cuckoo goes about finding them. I heard one calling from above me on a spring trek. Then it made a shallow quivering laboured flight down to the heather, and flew up to a low birch branch. I put the fieldglass on it, just in time to see it gulp down one of the huge brown hairy caterpillars. Then it quivered down again, found another, flew up to the branch and gulped that one back too. It seemed to fly against the wind, so it didn't have to beat hard or use its whole wing area – thus it could avoid startling its prey off its perch by a noisy arrival.

As we turned and trekked higher the female buzzard soared over our heads. She looked even bigger than usual, then I saw she was carrying something in her talons back beneath her tail. Suddenly she twisted to one side, and the object she was carrying sailed downwards nearer and nearer, then thumped down on the hill a few yards above us. It was a hefty sprig of dead brown heather. It was bad enough I couldn't find the nest she was apparently building; now she seemed to be deliberately bombing us!

One 28 April, after a long wet winter, we trekked over a 1,700-foot peak, down into the glen of the big river and up to the sanctuary I had come to call Stag Valley. Moobli was now used to tracking and began stepping high and scenting. We saw two large deer ahead. At first I thought they were large hinds but the hefty barrel bodies and bigger white rump flashes told me they were stags which had lost their antlers. As they carried on grazing, with Moobli now trained to lie down until called, I stalked them for photos, only to find a heat haze annoyingly blurred the images. Half a mile further up we had better luck with nine more stags. It was a late season, for four of them still had their antlers, while one had only dropped one. I wanted a skyline shot, so made my presence known. As they galloped away I took three photos, but as the lop-sided stag hit the skyline the camera jammed. I had run out of film!

On our way back we crossed the river by an old bridge made of square timbers bolted over larch trunks, and saw a sad sight. A deer calf had somehow got both its rear feet and one front foot caught up in the narrow gaps between the timbers, as if after getting its rear feet caught it had twisted round enough to paw its way back with the one front foot, only to get that caught too. In this tortured position it had perished – a horrible death. But two sexton beetles were happily mating on the carcass, the long neck and head of which were now moving up and down in the flow of the water with ghastly leers.

By early May the bluebells were carpeting the lower slopes, taking advantage of the light before the pale green lampposts of young bracken took over their areas completely. Higher up, trekking along steep bracken slopes in winter and early spring can be hazardous on dry days for the thick stems of the previous year's growth, golden brown, shiny, and lying flat in a downwards direction, can act like skis when you step on them. As your foot starts to move sideways the stalks snap off under it, gathering more bracken. If you don't immediately drop down you can be sliding downhill on a toboggan of bracken stalks.

I try not to judge any part of the wilds merely from its usefulness to man but I can find very little good to say for bracken. If young sheep, cattle or deer are foolish enough to eat it they get 'bracken staggers', or if they eat enough are poisoned. It spreads its pervasive underground rhizomes, unaffected by crofters' fires, over large areas, and helps prevent the regeneration of young trees as its tall ferny leaves block out the light. Flowers like wood anemones, sorrel, violets and bluebells have to get their seasonal cycles over before it is fully grown. About a million acres in Britain are infested with the plant whose toxins can be washed out and get into our river and lake water supplies. However, over the years I have found it does have some uses. When gathered dead and dry it made wonderful bedding for my wildcats, when burnt it produces more potash than most other plants. It does give some support to weak seedlings in the spring, especially wild raspberries, and when it grows round bramble bushes it not only hides some berries from birds but those that grow in its shade are often more luscious than the fruits above. It also gives shelter for some animals in summer storms. One creature that benefits by it is the larva of the lovely map-winged swift moth, which eats the roots.

Bracken can be helpful to the hill trekker in some ways. Its first short, stout massed green stems act as a check against the 'ski' slips I described, and it won't compete with tussock grass, so that before it's fully grown, and on the higher slopes where it is shorter, the ground below it is far more even than between the tussocks. That is a fact worth knowing when you need to husband energy on long hikes.

Coming back from one trek to find eyries of golden eagles we started zig-zagging up and down almost sheer slopes at the far end of the long wood three miles from home, until we came to a hooded crow's nest I knew was occupied in a tall ash. It was too high to see down into it, but the female was there, probably brooding newly hatched young, for as we hid behind a bush I heard her giving gentle croaks which were being answered by the male somewhere above. As we watched, he swooped in short circles round the tree tops then flew in with what looked like deer carrion in his beak to feed her. As he landed she greeted him with soft muted '*kar*' calls. Some moments later the female buzzard flew past overhead but, instead of a mobbing attack, the male flew away unobtrusively through the lower canopy of the trees and, when he was a good fifty yards away, came at the buzzard from a different direction. He was too cunning to call out or mob the buzzard from the nest itself as that might have given the nest away. The crow knows, possibly from its own nefarious proclivities, that buzzards will occasionally snatch a nestling, though not nearly as often as will the crow, which also purloins eggs.

Feeling the first midge bite on 16 May, I planted my entire vegetable garden in a long morning, then headed high into the hills with Moobli to break in some new boots. The first hog weeds were rearing lush jungle-like hoary leaves as if competing with the towering purple foxgloves whose velvety leaves lay back on the earth like green gloves. Near the woods greater stitchworts lifted their paired white petals on slender necks. The tawny grey winter look of the mountains was now changing from below, with swathes of green creeping up the slopes.

As we headed upwards the delicate traceries of pignut held their tiny white flowers aloft as if on invisible stems among the bluebells, a few buttercups had appeared, and the cerulean blue of the first speedwells, like giant forget-me-nots, formed ranks below small grassy shelves. In boggy patches the curled-edge leaves of butterworts, like saucy light green tongues stuck out by rude children, lay over the ground. A few had sent up their single lippy violet white-throated flowers on snaky stems, and when the breeze blew they swayed like tiny cobras. Among them were a few sundew plants, holding out long arms, at the end of which were round green leaves the shape of ping-pong bats, covered with sticky red hooks. Both butterworts and sundews are carnivorous plants: because they grow on peaty acid soils deficient in nutrients, they trap insects to supplement their diet. When a fly lands on the sticky butterwort leaf it folds over to enclose it and absorbs the juices from the decomposing body. The sundew lies like a sinister upturned red and green spider. Its hooks hold sticky drops to which insects adhere and its leaves too fold inwards.

After two hard miles of climb-walking we were up at 1,600 feet where huge grey rocks were embedded in the heather, and I realised my new boots were stiffer than I had hoped for I was already developing a blister on my right heel. We reached a natural 'armchair' in the rocks, Moobli panting only to cool himself but I from the exertion, and sat down. I shoved a double piece of brown paper between sock and flesh and scanned with the fieldglass the superb three-mile-long glen that stretched below us. Not a red deer in sight; just a few isolated groups of sheep like dots of cotton wool against the steep hills. But far down, flickering along on her hunting flight, was a female kestrel, her tail as redly barred as her upper parts and when the sun struck her back she shone like a coppery light.

A small flock of common gulls floated round one ridge of Guardian Mountain looking for caterpillars in the grasses, a pair of ravens soared south and a single hoody flapped lazily across the glen as if stroking the air with its wings. Suddenly there were two piercing '*kee keew*' calls behind us and as I turned my heart gave a surge. Flashing past above our heads were a pair of peregrine falcons. What incredible fliers.

The larger falcon was moving with jerky, strong, but slow beats of her scythe-like wings, as if flying contrapuntally to a weird music only she could hear, one beat to every two bars, while every other bird in the glen had to beat once to every bar, yet she was streaking along. The male or tiercel attending her did not fly this way but beat his wings two or three times to her once, yet he was moving all around her, up, by her side, down and up again. Then they both suddenly went into a power dive calling '*kee keew kee keew*' at a speed that was faster than falling. By the time I had the camera out of the pack and trained on them they were mere specks in the viewfinder.

At home later, nursing my sore feet in sheepskin moccasins, I heard a loud croaking, and hobbled out. The male hoody, watching from a high ridge look-out post as his mate brooded young in a tall thin pine in the west wood, was mobbing a raven and to my surprise the bigger corvid turned with loud protests and flew back to his own nesting cliff to the east. As the hoody flew back in triumph, the male buzzard appeared from the west. Now he chased the crow, whose pride was short-lived for now *he* had to twist-dodge downwards, and at great speed to escape into the thick conifers of the wood. Then the buzzard soared about in deliberate lazy circles as if reinforcing his right to be on the same territory. I was sure then the buzzards *were* nesting somewhere in the miles of lochside woods.

In glorious sunshine on 25 May Moobli and I were heading up the big river valley to check for eagle eyries in the flanking mountains at the far end, when we were treated to a bonus experience involving cuckoos and meadow pipits. By now the first pink and white heath orchids had appeared, with an uncommon early purple orchid among them. The four-petalled yellow flowers of tormentil rested coyly against the grasses, and amid the tiny cobalt blue chips of heath milkwort I found a rare white violet. Now the yellow spikes of bog asphodel were flaming the marshes lower down, just as they had carpeted the Elysian fields of paradise. Suddenly there was a slight thump ahead and a meadow pipit flew up, as if she had hit the ground with her wings on rising. I went over and in the grass under a large heather clump found her deep nest with four eggs, almost round and crimsony-brown with dark grey markings.

A few yards on I heard a loud cuckoo call and a male floated across in front of us quite slowly then landed on a long horizontal branch of a tall grey alder. I hissed Moobli back and we ducked behind a small bush. I heard the bubbling trill of a female, louder but like liquid pouring from a narrow-necked bottle, then saw her fly up from the ground forty yards away towards him. As she came near he also took flight and they flew round each other before landing back on the branch, close together. He bowed and twisted his long white-spotted grey tail sideways, first to the right, then to the left, and quivered his half spread blue-grey wings. He courted her with gentle '*coark*'s and odd growling noises.

Suddenly two meadow pipits flew up from the grasses near the nest we had found, clearly its owners. One fluttered towards the male cuckoo and the other towards the female, not attacking or buffeting with their wings but just hovering nearby, then going down to the ground by the nest and back up again, almost as if inviting the cuckoos to it. The meadow pipits did this once more, then one became more persistent and landed near the female cuckoo, but with no threatening darts or attempts to peck, such as are made when birds mob each other.

The branch was probably one of the female cuckoo's spy posts, from which she observed the movements of nesting birds, so saving herself a laborious search for nests in which to lay her eggs. Once a cuckoo starts to lay she has an egg to deposit roughly once every two days, up to a dozen eggs in all, so she tries to have a particular nest in mind. Whether she had known this nest before I had no way of knowing but she certainly did now! She just sat placidly on the branch, not responding to the excitement of the pipits, as if intent on memorising the nest's exact position. When the male cuckoo flew off the pipits followed him for a short distance. Instantly the female made a shallow quivering flight to the nest, pushed through the grasses, sat on it for a few seconds, then floated away again to join the male, both birds still being attended by the pipits.

Quickly I went over to the nest and found the cuckoo's egg, almost the same colours but slightly larger, beside the four of the pipits. How wonderful, I thought. I had never yet photographed a baby cuckoo ejecting other eggs or fledglings from a victim's nest and now it seemed, if I timed it right, I had the chance to do so. We walked away and had only gone twenty yards when I again heard the bubbling trilling call of the female. Then the two birds flew high over our heads, and I heard the weird wind-note '*hô hô, hah hah hah*' which cuckoos sometimes make when flying together. I often think it sounds as if they are having a good laugh at having planted an egg on the unfortunate foster parents, whose own young will perish and who will have to work hard to satisfy the appetite of the huge changeling as it grows to full size.

Knowing a cuckoo can hatch out in twelve days and will start demolishing the opposition within a few hours, I made a special hide of natural materials and a few days later boated out with Moobli and carried it back to near the nest, where I intended to leave it so the pipits would get used to it, until I could move it nearer bit by bit. But I found the nest had been ripped apart completely, pieces of the once neatly rounded bowl lay a yard or two away, and there was no sign of eggs or hatched young.

Suddenly Moobli put his nose down to the ground, blew it clear with a snort, and set off as if on a familiar scent.

'Track the foxy!' I encouraged him, sure that a fox would have been the culprit, and equally sure Moobli would not track it far for it would de-scent its feet in the river or a marshy tract before long. Yet Moobli wasn't behaving as if on a fox scent, no leg-cocking on tufts. He headed for the river, then turned left along one of its forks, and there ahead in the muddy-sandy patch were the tracks, just two, of a medium-sized wildcat. They were heading at right angles to a narrow water channel, so it seemed the animal had crossed it, after taking the pipit eggs or young.

Moobli lost the scent then but I zig-zagged over the grasses on the far side and searched the near bank of the main river. We didn't have to look long – for some twenty yards north of the original track line were three more of the wildcat's prints, going along the riverside this time but ending on a rocky shelf. We searched round again and a few yards past the outcrop were four more tracks turning towards a series of large rocks across the river, two of which were much smaller prints, clearly those of a kitten. It seemed the kit had not liked to get its feet damp and had been walking alongside through the grasses but had joined its mother when she had turned to cross the river using the rocks as stepping stones.

It was the year after I had released the last of the wildcats I had bred at Wildernesse, all of them to the south and west, many miles from the cottage. The previous season a shy kitten I had called Mia had cut loose on her own at the age of five months and we had not seen her again. Could this be her? If so, how wonderful to think that she had met a mate and now had a kitten of her own.

Chapter 17

The Mountains in Summer

Summer trekking for wildlife in the western Highlands is best done in June, before summer's height, when foliage is thick on the trees, bracken so high it obscures most animals and the mists and heat hazes also make photography difficult, even on the high tops. My main work in the first years involved the glorious golden eagles and also the red deer, but sometimes when going to or from eyries in the 300-square mile area round our home I could study both on the same trek. I also took a keen interest in the flowers.

On one eagle trek I had to go down a deep gorge and found so many fine flowers I almost forgot to scan the ledges above – luscious yellow globe flowers almost the size of tulips, water avens with dangling pink aconite heads, fleshy sea-green roseroot leaves topped by broad orange blooms and once, growing from a tiny pocket of earth in a rock crevice in the centre of the burn, a rare clustered alpine saxifrage.

Usually by late May the red deer are up on the high tops for the new grass and herbs, away from biting insects. If there are unusually long hot spells after a late cold winter the tops get 'burned off ', the bright sun and lack of rain killing much of the greenery as it comes through. The deer then have to go lower again. A wet May is better for the deer in fact. And in hot summer periods deer will make quick migrations down from the tops to eat lusher low herbage before dawn, then go up again in the first rays of the sun to where the winds are cool. Sometimes I have seen stags cooling off in the loch at midday, but only below wood-land into which they can rapidly escape.

It is during the last day or two of May and the first few in June that the red deer hinds give birth to their invariably single calf, in my area usually in the higher hill meadows. They wander distractedly away from the main herd, move in fits and starts, then lie down on one side, get up, lie down on the other, and occasionally as the calf is worked out front feet first, will give an odd muted bellow like that of an immature stag. Sometimes a hind will be standing at the moment of birth and the calf drops to the heather or grass with an audible thud. The hind then pulls away the foetal birth sac with her teeth and swallows some or all of it. She stays by the calf which after a few minutes struggles to its feet and instinctively searches, with frequent staggers, for its mother's teats. The hind stays with the calf the first day but as it grows stronger, she leaves it lying for long periods in the grass or among heather or bracken to go grazing. She may even wander as far as a mile away. Then, remembering where the calf is, she comes back in the afternoon, gives a soft bleat and up gets the calf for more milk.

Due mainly to intensive eagle work, I never had time to search for newborn calves. One early June day in the sixth season I decided to take a rest from eagles and instead trek up to the 1,800-foot meadows, hoping for the luck of the man who broke the bank and maybe to photograph a hind actually giving birth. What I did get, after descending a steep ridge, were some unusual pictures of hinds in the height of the early summer moult. They were a dishevelled mob, with the red summer hair pushing out the grey winter coats along their tattered flanks. With large whitish areas on their rumps where their old alarm patches were falling out, they looked like giant roe deer.

I was just wondering why none of them had calves when suddenly, in a small niche in the grasses, head tucked between its little front hooves hoping it wouldn't be seen, there was a lovely dappled cream-spotted calf. It was a mere twelve feet from me, and with the wind the wrong way Moobli had neither seen nor scented it. I took four pictures as it shrank there. Then, fixing my eyes on a spot a foot away so they wouldn't meet the calf s gaze directly, I stole nearer, fully expecting it to get up and run. But it didn't, and I managed gently to hold it round the neck. Immediately it let out a piercing '*bleea!*' so startling I nearly let go. Its little grey nostrils were dilating, its beautifully long lashes blinking slowly and fearfully over its huge eyes. Moving very slowly, I sat down and talked soothingly to it, stroking gently as it laid its head and neck across my legs and relaxed almost totally. I took another close photo, then got up carefully and let it go. It staggered away a few steps then turned as if coming back to me. Moobli, intrigued too, moved forward to sniff it gingerly and the calf ran off, very unsteady on its feet. I doubt it was two days old.

I caught another one in the same way, moving very slowly, Moobli keeping well back, but again resisted the impulse to take the beautiful creature home, knowing well its mother would be back to feed it in the afternoon. On the way back we found two more, but I just took photos from a short distance. When I later told a local keeper of the day's experience he was surprised that I had caught the calves without using a dog or a calf-tagging net.

'You lost yourself £200,' he laughed. 'The deer farm is paying £50 a calf this year!'

I replied I wasn't *that* hard up, and while I hoped to make a meagre living from nature writings and photos I didn't want to turn wildlife into a hard business. As always, nature in such circumstances is better left alone. There is no excuse for believing lone deer calves to be abandoned – unless you have spent at least thirty hours watching one, and from a place where the hind is not scared of returning because of your scent, or you know the mother has been killed.

Unlike domestic sheep which on the same hills frequently suffer complications when lambing, red deer have very few casualties, of mothers or calves. We came upon a tragic little scene one summer afternoon as we headed up a high mountain having just photographed a young cuckoo being attended by *four* meadow pipits. Moobli suddenly thrust his nose into the ground and began sniffing noisily. I went over to find what looked like a fox hole. I bent down and choked on a bad musty smell. It seemed there was a dead fox some two feet down for I could see dark red brown fur below, seemingly stuck in a peak runnel. Seeing what looked like a leg, and overcoming distaste, I thrust my arm down and hauled out a dead red deer calf. As I looked at the poor long-legged mud-covered mite lying on the grass, I imagined how it had stumbled into the runnel, made in the peat by winter rains, and hadn't the strength to climb out again. With her cloven feet its mother had been unable to help either.

I learned early in hill trekking that these runnels, often three or four feet deep, can be dangerous to man too. Once I put my foot into one hidden by long *molinia* grasses, and found myself waist deep in peaty ooze. Another time my foot went two feet into a hole that held the leg rigid but pitched me forward. Had this tight hole had a sideways angle, pitching me to the side, I would have broken my leg. These runnels are always dangerous when travelling at speed. Whenever I hear water running under-ground in the hills I am very careful where I put my feet.

By mid-June our mountain treks produce more treasures – a small patch of uncommon dwarf cornel, the purple-black centres of the white squarish flowers staring up at us like eyes. Small pearl-bordered fritillary butterflies dance like orange flames over the hillside, with here and there a larger dark green fritillary, winging more powerfully, looking for the nectar of the first knap-weeds and thistle. In the grasses red and green grasshoppers saw their jagged legs to make their fiddling songs.

Once, at about 1,300 feet, we stalked three hinds and a young stag, his two spike antlers clad in velvet. He actually came forward to have a closer look at the two crouching objects. Coming over a small ridge, I almost stepped on a grazing hind which, with a sharp bark, took off to the west. Suddenly a two-week-old calf with masses of white spots on its dark red coat, which had been lying in a cleft, also shot up and away. I was astonished at its speed – like a big jack rabbit, sometimes travelling with four-footed jumps like a springbok. It gave piercingly loud squeals as it went, as if to call its mother back.

On the way down, through a marshy area, and among the white and early purple orchids, I found two lesser butterfly orchids. These yellow spiky flowers with long green stamens are quite rare in the Highlands, being usually found on the chalkier soils of southern Britain.

If it drizzles when you trek at this time of year the bracken, often head high, the fully grown tussock grass crowns, and the tall cow parsleys and hogweeds all hold water like sponges. Most of the time you're soaked from the waist down, but to wear rubberised shower gear in summer is to become a walking Turkish bath.

Living near a loch, the little biting midges are now at their most numerous, diabolically programmed to locate you for a mite of blood to fertilise their offspring. Once, when trying to transplant three white bluebells into a fenced area, there were so many the buzz of their tiny wings made a high-pitched zinging sound and, with my eyes inflamed by itchy bites, I just shoved them all into one hole and ran! Scything down an acre of bracken several times each summer – until in the sixth year it was almost finally defeated – was a test of torture and no spray seemed effective longer than twenty minutes. At this time in July there is a two- or three-week period when the slow zoom-flying deer flies, known as clegs, compete with the midges, airily dodging your swats and landing back in the same place for another try, their bites like the thrust of darning needles. No wonder, because of these and tabanids and bot flies, the deer go to the windier tops in summer.

Trekking in summer is a sweaty business and perspiration is the last thing you need when stalking deer or any mammals which scent it fast in the humid, often misty air. My old Canadian method of boiling spruce or pine needles to deodorise my stalking gear became tedious but I found that bog myrtle, also known as sweet gale, which grew on the boggy flats near my home and filled the air with perfume, was as effective and far easier. As the Indians used buffalo grass when hunting so I now crushed handfuls of leaves, mixed the mush with a little water, then spread this over face, hands and the areas of clothes most prone to perspiration.

I'll never forget the day – 20 July – I first used it. It was unusually mistless, with sun and cloud alternating, and a nice cool breeze. I plugged upwards, glaring at the ground, my legs working like a disembodied engine that seemed no part of me, until almost reaching the 1,500-foot peak overlooking the long glen south of Guardian Mountain. I had seen no deer yet and Moobli was receiving no scent in the south-west wind. As we headed east to north below the peak (never show yourself on any skyline) I saw a hind lying down fifty yards up ahead, and standing beside her was her young calf. I was afraid she'd get our scent in the cross wind but through the fieldglass I saw the breeze was being deflected downwards by a large knoll just to the west of her. I backed down, hissed Moobli to keep back, then crouch-stalked north to a small heathery prominence. As I took photos the hind just kept chewing the cud, jaws moving from side to side like some gormless gum-chewing yokel, her glassy hazel eyes appearing to look right through me. Then she casually got up and ambled to join two other hinds with calves which had appeared over the brow from the northwest.

I was about to move back and upwards again to get nearer the group when further away a hind with two fawns at her heels came upwards from our left. I fully expected another hind or two to be with her but when none came I suddenly realised these could just be twin calves, extremely rare among red deer. Heart thudding, motioning Moobli to lie down near the pack, I slid like a snake through the summer grasses and round tiny prominences until I was near enough for really good pictures.

Now came magic moments. The fawns, smaller than the others we'd seen, still had lovely white spots on their red-brown coats. They nuzzled each other, one chased the other on to a small hillock, they butted heads together, then one reared up on to its mother's back as if trying to mount her. As my camera clicked away, I whispered 'More light please' and lo, the clouds scudded away from the sun as if obeying my stage directions, bathing the dappled fawns, light red coat of the hind, the grey rocks, green grasses and flowers in beautiful light. Then this light dimmed to softness, to shade as more clouds passed by. Then came a few beams, followed by stark bright light – every nuance I could have hoped for. I was now certain they were twin calves for both were familiar with the hind and there wasn't another deer in sight, for the others had gone over the ridge.

Some high summer treks were abortive for deer, despite my new bog myrtle discovery, because there was little cover on the high tops and the light helped their sight. If the deer didn't spot you, then the occasional raven would – and if it flew over the stags or hinds with a loud '*kwork kwork*' the deer took heed, became more wary and located you quicker than they would have normally.

But there was always some little bonus – the first purple scabious out on the lower slopes in early August, or a rare patch of the green-veined white flowers of the lovely Grass of Parnassus. As we came down the lower slopes one day, fording our way through tall flowing tussocks, floppy dark brown Scotch Argus butterflies barely bothered to get out of our way. They flitted on chocolate wings between the yellow hawkbits and occasional fat-stemmed hogweeds, over whose white multi-flowered heads ran little hoverflies and bluebottles, relegated from the company of bees and butterflies whose longer tongues allow them to work the more exotic flowers.

How often on these summer treks were we made aware that this is the season of the young, apart from deer calves. Baby meadow pipits sprang from almost under our feet on the hills, and on the lower slopes where bracken grew to six feet high small flocks of young stonechats, attended by their parents giving anxious '*chak chak*' calls, flew from frond to frond just ahead of us, and wren families whirred like large brown bees, the parents chirring coarse alarm notes at the even more stumpy-tailed fledglings they were teaching to find their own food.

As we neared the top edge of the long wood a mile from our home on one trek in early June, I heard the piercing clarion call of one of the buzzards just behind us. With the least movement I removed the cap from the telephoto lens, turned and raised the camera. The female buzzard was right above our heads, the late sun gleaming on the beautiful colours of her underwings, and I got the best flight picture I had so far taken of these birds. As she beat onwards with the faintest woofing of her pinions, I realised that if it had taken me seven Highland years to get such a picture, it was even more incompetent to have found only three unused nests, never one where I could watch and photograph the youngsters. That day I consoled myself: I had at least spent the last two seasons working with golden eagles, and therefore hadn't had time to concentrate on the 'lowlier' buzzard. Oddly, it was only two days later that my chance to work with these buzzards did come – and mainly because of eagles!

I was boating up the loch for the last stint with the eagles, and to take the hide down, when the female buzzard flew across the loch, over my head, then glided up into a giant old oak tree which stood out from the smaller trees round it a hundred yards above the shore. Next day, after the usual weary eagle night out on a cliff ledge, the dismantled hide now in the boat, I was dawdling back, looking forward to a few drams, supper and bed, when I remembered the buzzard. I didn't feel like climbing up yet more rough steep land amid the long bracken and snaggy rocks but I told myself 'Duty calls,' went ashore and climbed and clawed my way through the wood up to the huge oak.

There *was* a buzzards' nest in it, in a large four-branched crotch forty feet from the ground. Passing the tree, noting white splashes round its base, I climbed up further to a tree-lined ledge so that I could see into the nest. There was an almost fully fledged buzzard chick, like a small eaglet, in it too! The ledge was ideal for a hide, to be made between two small rowans, exactly opposite the only open spot where the parent birds could fly into the nest. The chick crouched down but watched every move I made with dark brown eyes.

Exhausted though I was, excitement restored energy to me. I climbed back down and, sweating worse than before, hauled the hide, its hazel poles, a cushion and plastic ground sheet back up the slopes into the wood and hid them well out of sight of the nest tree under a covering of bracken.

Next day I filled two sacks with heather and leafy green bracken to obscure the hide, as I didn't wish to upset the parent buzzards by tearing up vegetation near the nest, and carried them to where I had left the other materials. There, out of sight of the oak, I partially built the screen-hide, rolled it up and, strapping the now hefty bulky heathery object to my pack frame, carried it to where I could watch the nest tree through the fieldglass.

I didn't have to wait long. Within half an hour the female soared into the tree, surely with some food for the chick, then flew out again after two minutes. As she winged away eastwards I saw with relief that she was joined by the smaller male. I reckoned I had at least one, maybe two hours to get the hide into place without being seen by the parents, for with such a well-fledged chick they would leave it alone for long periods. After many hot struggles I had the hide in place, tied with green gardener's cord to the two rowans, some branches of which were woven across the front and top. Tall bracken grew up in front of the mossy ledge and I weaved some of this into the hide's front too. A few leafy hazel branches completed the camouflage and as I left I noted with satisfaction the hide was indistinguishable from the thick herbage round it. I took a couple of photos of the chick, now standing bolt upright in the nest and watching me with great interest, and left.

On 5 July, with a hunk of bread, lump of cheese, an onion, a slice of fruitcake, can of orange juice, camera and tripod, I trekked back to the hide in late afternoon. (I dared not leave the boat on the shore below the nest tree.) Poor old Moobli had to stay behind in the cottage. It was a hot, sultry and hazy day with that dense close white sky denoting the possibility of thunder, and the sun blazed down like a fiery lemon. Still, anything was better than rain, when good photos are impossible, and I was afraid the recent dry spell might be coming to an end.

There was no sign of the adults as I climb-walked in. I hastened my final clamberings, thrust my pack into the hide, squeezed in beside it and closed the herbage-stuffed flap. I was now in a three-foot by four-foot cubicle of dank vegetation where it was impossible to lie down, and the only alternative to sitting straight on buttocks with knees drawn up below my chin during the numbing night hours ahead was to switch to either left or right haunch. Either action had to be made with the utmost slowness and care so no sound or sense of movement could be perceived outside the hide.

To my surprise, no sooner was I in than the buzzard chick stood up straight and started a high cheeping call. Then the female buzzard swept in, a flash of tawny speckled brown, and landed on the nest by the chick which was now semi-prostrate with wings open as it cheeped with beak down for food. She had brought none and seemed to be just checking the chick was all right. I froze, enduring the first midge bites, cursing that I couldn't take a picture for there had been no time to set up the camera. At last she turned and flew but then quickly landed on a bare snag to the right of the hide, facing it and appearing to glare in. She began to call loudly, piercing 'keeyoo's which made my ears ring, then I heard her mate answering from afar, his calls becoming louder as he came nearer.

I was now suffering torture from the midges as hordes of the little pests scented my presence and zeroed in on to every patch of bare skin. Not daring to move, my silent curses increased as I realised I had forgotten the insect repellent. Finally the female buzzard flew away, landed in another tree out of sight, gave two softer calls which made the chick squat down, then flapped away.

Quickly I got the camera on the tripod, thrust the lens through a hole at eye height and stuffed bracken round to hide it. I now saw the adults had decorated the nest with leafy green birch twigs, and that the chick was virtually full grown, with a full-length tail and almost no white down left. No sooner was the camera in place than both birds returned, the female landing on the bare snag near the hide again while the male perched on a branch on the far side of the nest, hidden by thick foliage. Again, no food was brought but the female still called from time to time. As dusk drew on the calls lessened in power – just a drawn out '*kairoo*' rather than the metallic high '*kee-oo*'.

After a long silence I thought they had gone again. I munched my frugal supper and tried to deal with the midges. I wiped onion over my face and neck but it didn't keep them off. Raw orange juice was tried next but was just as useless. At least my hands were protected as I was wearing gloves, a trick I had learned from working the eagles, but the midges worked under my cuffs, down my neck and covered my face. Finally I tried to endure them, gently rubbing them to death with the gloves when I could so without making any noise. When I took the top off the can of orange juice I realised the female was still on the snag and that the slight pop must have woken her up. She gave a soft croaking '*kaoo-urr*', as if the normal call had been slowed down on a record player at the wrong speed.

It seemed buzzards came in to roost much earlier than eagles and that now she would be on that snag all night, near her nest but out of camera view. I was disappointed for I had wanted to see her on the nest with the chick for a long period. Now it seemed I would have to endure the night perched on one numb edge of my backside or the other, the midge bites, the fetid sultry heat, the whole masochistic yet creative business of such work, all for nothing.

Then just after 11 p.m. in the near dark there was a flapping of wings and she flew to the nest. The chick immediately tunnelled and pushed itself under her, between her widely straddled pantalooned legs with little snickering squeaks. The buzzard seemed a good mother and despite her own discomfort at having such a lump under her, brooded the chick all night. She actually hooked the youngster inwards with her beak, as if stuffing it under herself and, though it was too dark, I took a few photos. The only slight alleviation to my midge torment was to tie a mohair scarf round my neck and face, hellish ticklish in its own right, almost suffocating in the heat of what was certainly the hottest night of the year. The hours passed . . .

In the dark twilight before dawn the chick seemed hungry and, with its mother now standing to one side, called harshly '*kaoww*', almost like a crow – a sound which it continued to make when the mother flew off and which I soon learned was associated with hunger. When the sun came up it appeared as a huge red ball, so the light was still not good. At 4.30 a.m. the mother swept back in and landed with a vole, which the chick tore up easily, standing on it and rending upwards with its beak. The vole was gone in three swallows. An hour later she came in again, flying with perfect silence: if I had not been looking through the lens at that moment I would have missed her. She had a frog in her talons. She turned to go almost immediately, for an irresistible shot coming straight at the lens, and wheeled away. I thought she had gone but as I eased my cramped position I heard her call from the bare snag near the hide. It was the soft '*kairoo*' call again and it made the chick stop hauling at the frog and freeze perfectly motionless, hunched down like a broody pullet, its eyes blinking.

After a silence, the chick stood up and pulled at one of the frog's legs again but the mother looked over, '*kairoo kairoo*', making the chick crouch and freeze once more. It seemed she was suspicious of something, but I was sure it wasn't the hide or she would not have perched so close to it. Then I heard a rustling below: a roebuck was walking past the oak tree, negotiating the rough rocky ground with care as if suffering from arthritis of the legs. One sound from me and it would have bounded away, as fleet of foot as ever, and I daren't shift the camera to photograph it for fear of scaring the mother buzzard. It was interesting that the mother buzzard could make the chick freeze with these odd soft calls. If she used them when a human was approaching the chick would crouch, and a man looking up would think the nest was empty.

When the mother flew away the chick resumed eating but didn't appear to like the pale orangy meat of the frog as much as the vole. Its meal complete, the chick backed to the edge of the nest and squirted its white faeces well clear. As the sun rose its light became yellow-white and gleamed into the nest but the big branches and foliage threw shadows everywhere, making good photography difficult. At about 6.30 a.m. the two buzzards flew over the nest tree, crying to each other, his cry shorter and higher-pitched, hers longer and sounding almost cracked. Then their calls faded into the distance.

A mere twenty minutes later the male, a smaller neater-looking bird, came in and dropped half a frog, having doubtless eaten the rest himself. It seemed clear buzzards feed their chicks mostly in the early mornings, which made sense for the hunting of small creatures is best around dawn and just after. Buzzards, I now realised, were really just small eagles. They occupy a slightly different ecological niche, using the lower slopes and also hunting in the more open woodlands, taking some prey, like beetles, larger insects, frogs or snakes, that eagles would not, or only rarely. I think the name 'buzzard' implies something vulture-like which scavenges for decomposing flesh, yet this fine bird, while eating fresh carrion like many birds of prey, is mostly a clean hunter. I feel that Hill or Woodland Eagle might be better names.

After eating the frog the chick stood on a thick stick and tore fibres out with its beak, as if strengthening its neck muscles, then held it up in the air like a weight lifter. It flapped blunt wings, not yet fully grown, bouncing up and down on the nest as if rehearsing for the first take-off. Then for a time it dozed on one leg, its crop swelling full. After a rest it flap-climbed up one of the four sloping branches of the fork, rearing upwards as if going right to its top, then flopped awkwardly back on to the nest sticks, flapping hard. Occasionally it tore off corky bark from the surrounding branches, maybe to gets its beak strong.

At 7.45 a.m. the mother glided in, closing her wings to form an arrow as she shot between the branches, and landed with a woodpigeon squab from which she appeared to have eaten the wings, legs and head. These two parents were excellent hunters and I was glad they had not noticed the hide. When she had gone the chick walked to the young pigeon and with a hard fast strike hit the body with its right foot, sinking in the talons as if to 'kill' it, and it often squeaked between rending the small portions of meat.

I waited a good while longer, then with the sun high and no further sign of the parents, the chick asleep with its head resting on the sticks, the heat increasing and the midges still tantalising, I decided to go.

Although I had spent only seventeen hours in the hide, enough was enough. Next time I'd be sure and take my 'midge juice'.

As I walked back through my woods a little flock of long tailed tits flitted past me, filling the air with high *'tsee tsee tsee'* notes, clutching little twigs as if for grim death. When I opened the door Moobli shot out, buried his head in my groin, whining with pleasure, but as usual not barking. Tired as I was I had to throw his wretched great sticks until, after a quarter of an hour of hard pounding up and down the north hill, he was panting like the bellows of an old steel furnace.

I could not get back to the buzzard nest until 14 July. I climbed the hill, bottle of midge juice banging in my jacket pocket, and had a shock. The nest was empty. The chick had flown. I was disappointed that day but was lucky enough to see it several times, both alone and flying with its mother, throughout that summer.

Once, we went for an early morning stroll in the east wood when the young buzzard flew right over us and landed on a dead larch branch and called a ringing '*ping-oo*', though not quite up to its parents' standards! As it was in the wood again a few days later I felt sure it hadn't learned to fly really high yet but was instinctively looking for a territory, perhaps at first within the large territory of its parents, which seemed to cover a good six miles of the loch shore woods and hills. Two weeks later I saw both chick and mother together, flying over their nesting woods. It was a cheering sight, and now I knew the chick was female, for she was the same size as her mum!

Tired of writing indoors during a few days of rain, we braved more drizzle showers on 4 August for a long trek up and over the great dished corrie to the north-east, over the 1,700-foot peak and down along the long river valley. The burn was raging with white water over knee height. I figured there was no need to get soaked right at the start, so I scrambled across the upper branch of the twin beech fork that had fallen across it. Even though I was on hands and knees I almost fell off. Moobli spurned such prissy tactics, worked out where the shallowest parts were and hurled himself at the burn at a run, sploshing through and going diagonally downwards, his rear section seeming in danger of being swept away as he hooked himself momentarily on to rocks with the great non-slip paws and claws of his front feet.

As we crossed the upper ridge of the corrie I saw a bump on a rock I had not noticed before some 400 yards to the north. Through the fieldglass I saw it was one of the peregrine falcons. I took the camera from the pack but too late: the falcon hopped down into the grass on the far side, just as its mate swept behind it, and both were lost to view behind the ridge.

Another discovery that day was that midges were biting at 1,700 feet in the humid air – when I had always understood they never ventured above the 1,500-foot mark.

On the long trek up the river valley we found a large herd of stags placidly grazing near some thirty hinds and calves. They were displaying no rutting behaviour whatever; their antlers were still in full velvet which in a few days would start to be shed. It was a showery day, cold for summer, and the hinds had probably come down from the tops, just happening to be on the lower more heathery ground which stags prefer, because of the weather.

On 2 September we were over 2,000 feet on the south slopes of Guardian Mountain when we found a huge stag, his antlers now shorn of velvet, with ten points, grazing with a herd of thirty-two hinds and calves. He was showing a proprietary attitude towards them, the earliest I have recorded a business-like stag with hinds, though he was not yet roaring.

By mid-September the hills start their most spectacular colour change of the year. The blackberries have ripened on the lower brambles, and higher up the tussock grasses are growing brown at the tips on the more exposed slopes, with the stems turning orange lower down. This deep orange works its way down to the bases so that the hills seem to be licked by flames in strong breezes. Purple scabious is now the lord of the mountain flowers, the latest to survive, with occasionally a patch of Grass of Parnassus still holding up a few white blooms.

I was bending to photograph some of this grass one day when I heard a noise of rushing air. Two ravens had spotted us and knowing this to be the deer stalking season, were hoping I might have left a gralloch. They dived clown close, the wind roaring loudly through their black feathers. In the woods and along burns where there are a few trees, the quaint little deer keds with their squat brown bodies and tough little bow legs are swarming, ready to hone on to the heat and sight of any passing deer – and when they land on one they shed their wings. But they hit into dogs and humans too, and if they don't realise their mistake quickly and buzz off again, they lose their wings – their finest form of transport – and will die without ever finding a deer.

Around the third week in September the first hailstorms fall upon the land, striking off leaves and hissing in the undergrowth. I saw the young buzzard once, flying slowly overhead. Suddenly it crinked back its head as if it had been hit on the neck by an extra large hailstone, and flew on. It had learned much about hunting by now and its relations with its parents had become loose. These first hailstorms seem to act as a trigger upon the red deer up in the hills, for it is about now the master stags break free from the bachelor herds of the previous ten months and travel many miles, usually back to favoured places, to find hinds and round up as many as they can into their 'harems' for the rut. The earliest I have heard stags roaring was on 19 September in our third year.

We had trekked over the high crests to the huge corrie east of Guardian Mountain and hadn't seen a single deer. Then, as we came into sight of the corrie, rounding a high rock face, there ahead lay the largest herd of hinds, calves and yearlings I had so far seen. Some of them had already crossed over a small ridge but I counted 152 deer. There were two stags with them, one a large dark grey beast while the other, much younger, was a light fawny colour. The rut it seemed had not yet started in earnest for they walked along quite peacefully a few yards apart amid the hinds. Occasionally the bigger stag walked towards a hind as if to sniff her, but each hind promptly trotted away. We stalked them for a few photos but it was a bad situation – the light was poor, they were too far away, and there was nothing between us and them but an open valley.

As we left, the big stag began to roar. He walked to a small mound and with swelled neck and black mane thrust out, he bellowed his first challenge across the green and golden hills. The younger stag moved away.

Soon the rest of the master stags would come in from the surrounding heather hills and the big stag, which had clearly just arrived among the hinds himself, would be hard put to secure as many of them as possible into his own harem. Within a fortnight this herd would be broken up into a dozen or so groups, the biggest and fittest stags would split their harems away from the others, and not tolerate any rival trying to steal hinds away, or even come near them. The red deer rut, the most exciting time for the hill-trekking naturalist, had begun.

Chapter 18

The Mountains in Autumn

We were heading north up the steep slopes on a fine early October day, pausing at every crest to look ahead, and as we furtively looked round the last rocky ridge above the great dished corrie the noise hit our ears in a flood. The hills were filled with roarings – the guttural bellowed challenges of the master stags, the short deep grunts as one stag rounded up a straying hind, and the occasional higher-pitched blatting barks of younger stags which also wanted hinds yet daren't approach the harems of the masters while they were still in the first flush of the rut.

At first my eyes could not pick out any deer for all the groups were scattered over a mile away, their coats perfectly camouflaged, yet the vast amphitheatre of hills amplified the sound and if I had known nothing of red deer I could almost have believed we were surrounded by unseen lions. Searching towards each roaring with the fieldglass, I picked out one by one the different groups and tried to decide which to stalk first so that, even if they ran after the final approach, they wouldn't alarm the deer ahead or move across the great shallow corrie for all to see. Ideally, I wanted a group I could stalk and photograph without them knowing.

Suddenly there came a loud anguished roaring high to our right and north-east of us. I peered between the rocks and saw a young red-coated stag half a mile away on an exposed ridge with a few hinds below him. We ducked back to the south, headed east so that the ridges would hide him and began to stalk upwards. It was my 970th Highland trek in ten years and on the south-south-east wind we were taking chances, but I thought I knew the hills by now. Wind is to air what water is to oil, it exerts a downward pressure; and there were three undulating valleys topped by small crests which sloped up to the north-west between us and the stag. I hoped our scent on this worse than a mere cross wind would be deflected by each of the ridges, and carried away from him.

We took our time but when we reached the last ridge it seemed I had been wrong for he had gone. Prints in the mushy turf after morning rain showed that stag and hinds had not run but had probably just moved over the skyline to the east. I knew better than to go after them and so expose ourselves. Then we heard deeper roars to the north. A master. Another half mile – and there was a huge black stag with thirteen hinds and six calves on a high grassy spur north of a small but deep gully.

It was a hard stalk. I crawled sideways like a crab so as to present the herd with the smallest possible view, Moobli creeping along as my shadow behind, over two open turfy areas with just a few rocks for cover, then through a peat bog channel and steeply upwards out of it towards two huge slabs of embedded granite. We were now within 200 yards. The rocks were too high for me to lie behind and still see through the camera and too low to kneel behind. I spent the next hour in agonisingly cramped and twisted positions, watching and taking photos. Not once did the stag try to eat but kept moving about, standing often to roar. If any of the grazing hinds strayed too far from its imagined harem circle, the stag trotted towards them with head extended, making short grunting `*but but but*' noises to round them back in closer. Once it quite viciously chased a calf which had moved too far to the east, snapping teeth at its rump.

The only other males in the group tolerated by the stag were among the young calves of the year, and even the yearling `knobbers', their pre-antlers mere hair-covered knobs, were driven from their mothers for the first time in their lives. I saw two grazing miserably about a hundred yards from the harem, casting reproachful glances towards their mothers, not daring to go too close, knowing the master stag might drive them off again.

Then I saw the master sniff a standing hind as if thinking she was ready to mate. When she trotted smartly away he ran after her with his tongue out, trying to lick her rear, with a lips-raised leering look. Now and again he stationed himself next to a sitting hind and nuzzled her neck, and two of them nuzzled him back, one licking his cheek as if fondly, but none of them got up. Each time a hind slipped away from him he would give up after a brief chase, then pause to roar, his voice trailing off as if with frustration, and they would stop and graze again. Monarch of the glen? I thought. The hinds basically please themselves, grazing more or less how and when they wish, and the stag has to wait until *they* are ready to mate, which is usually only one day in twenty. And it is nearly always the hinds which are most alert to danger. If you have stalked badly, it is a hind that scents or spots you, gives a short hoffing bark, and away goes the herd while the amorous stag, mind intent on other matters, gazes after them with surprise. Then he reluctantly trots after them.

A higher more bleating roar sounded above to our right. A young stag with six points had ambled over the crest with just five hinds, and I took a few photos. When the youngster approached one of the hinds as if to mate she also trotted away, then kept on going towards the big stag and the larger hind group. The young stag watched her heading towards his rival's herd but thought better of trying to round her back for it would have brought him too close to the master.

I changed the film and, wanting closer shots of the big beast, slid backwards out of sight to the south-west, then stalked upwards to the north, slowly climbing up to where the master stag would be, keeping slightly leeward of a big rock I had fixed in my mind as a reference point. Moobli knew his duty so well now that orders weren't necessary, just a brief 'sit down' motion with my fingers and a 'good boy' pat. I could hear the roars loudly now and my heart was pounding as I slowly peered through the grasses on to his ridge – the master was only forty-five yards away and wouldn't all fit into the picture!

I backed off ten yards to the left and came up again. He was now trotting to round up a hind with short grunts, looking like a huge horned dog, his whole body rippling with muscles. I could hear the pounding *tum tum tum* of his piston hooves on the short cropped turf of the hill. Despite the hundreds of times I had now been close to wild stags in these hard jagged hills, the thought 'What if he should go for me?' again came to my mind. I would have had little chance without a gun. Maybe Moobli's fierce barking plus my frantic jacket waving would scare him off. But at these times you are acutely aware that fatal attacks on man, though rare, *have* occurred. I took a few more photos of the master roaring, shrank back down and sneaked northwards below his ridge.

You have only done your job well as a photo-stalker if you get your pictures without the stags or hinds ever knowing you are there. We got ten more stag harems in the camera lens in a fifteen-mile trek that day, but it was not always so. In the early years when Moobli was a pup, my own inexperience and the difficulties while training him produced a few abortive treks. The last of these occurred a month before Moobli was two years old.

The day dawned in a bright lemon sky which turned to deep blue as the sun rose, but a south-east gale was blowing. I decided I had little choice but to go with the winds to the high crests south of the western end of the glen below Guardian Mountain, cut over the tops, then work back eastwards, slightly against what would be a cross wind.

Once at the crests, we crouched behind some rocks and heard and saw a young stag with four hinds roaring away on the far hill to the north-east. We went a bit further, then saw below this group a huge mature stag, weighing a good 300 lbs, which had seventeen hinds and some calves in his harem. They were halfway up the far hill too; we were high and there was nothing between us and the two groups but the slopes below us and the deep valley. So how to stalk?

We first struck north-west down a deep gully, in the shade of the mountain tops above us, making it hard for the deer to see us when looking against the sun to the south. Moving in gullies and from rock to rock with aching slowness and all hunched up in the open spaces so the deer would not distinguish a human form if they saw us at all, we managed to reach a tussocky ridge some 300 yards from them. The sun was on the land just ahead and, to avoid it, I dropped into deep peat trenches and with painful knees crouch-waddled like a duck over the soft blackness for the next fifty yards. Then, privileged spectators in a magnificent wildlife tableau, Moobli lying beside me this time and even keeping his great ears down, we settled down to watch and take pictures.

This big stag had a fair but lopsided head with eleven points and, like all masters, seemed constantly on edge, walking round the hinds, chivvying the outer ones from their grazing for no reason it seemed except that as he couldn't relax why should they! Occasionally he thrashed his antlers against heather or high tussocks sending spikefuls of herbage high into the air, as if to demonstrate his strength. When two hinds wandered down towards the river he glared, ran like a trotting pony, his head and neck level and antlers down over his back, making short 'uh uh uh' grunts as he rounded them back up.

After half an hour two younger stags, about four and five years old, with smaller six-point antlers came downwind over the hill from the north-east but stopped a quarter mile from the big stag's harem. They grazed little but seemed fascinated by what they saw, staring at the hinds as if longing to move in and take a few but not daring to risk the master's anger. They looked like students, old enough to feel the urge but knowing they weren't yet tough enough to mix it with the big boys.

The smaller stag above the main group, possibly a six-year-old with eight points, just walked among his four hinds, not chivvying them about like the big stag. Occasionally he took a few paces towards the master below and bellowed a challenge in a higher timbre voice. But as soon as the big fellow took a few steps upward, he wheeled abruptly and went back to his own hinds, not wanting close contact. After twenty more minutes four hinds came up from the river, placidly grazing. The master saw them and ran down the intervening 200 yards to try and herd them back to his main flock, but apart from wheeling away quickly when he got close they refused to go. He could not have felt too determined for he just seemed to give up and walked back to his harem, then traded roars with one of the two young stags who had come closer after seeing the master go downhill.

Hoping I might get some fight shots I decided to chance stalking further over the rough but open ground, and telling Moobli to lie down, slithered to within ninety yards of the main group. As I peered through the grass on a small crest, I saw to my disappointment the hinds running away. Moobli had decided to follow me and was walking up the crest, his tan chest blazing in the sunlight, and they'd seen him. I hissed him to lie down and gripped his neck ruff, whereupon he gave a loud yipe and the hinds below also bolted. Oddly the stag hung back, turning to each hind as it ran past as if questioning their departure. But they were off! He had no control over them at all once they saw or scented danger. I got four more photos before the stag trotted off in their wake, looking most put out, still not knowing we were there.

Moobli's moving had been partly my fault, for I had not looked back after the first yards and hissed the usual 'Stay there,' but I'm glad to say that was only the third – and last – time Moobli disturbed a deer stalk.

I found out in the second year that if a cold north-east or easterly wind dominated a two-week period the deer tended to gather on the far westerly slopes to escape it. And if the movements of the various estates' stalkers coincided in one area they thought nothing of a three-mile move overnight to less dangerous ground. As the shooting season advanced they became increasingly wary of any moving human. I restricted our treks at this time of year to Sundays because the estate stalkers have a hard task to stalk and select the right ageing and poorer beasts to be culled by their rifle-toting paying guests without walkers disturbing the herds.

The main hunting is done in the first three weeks of October; three days in three weeks is good enough if you know what you're doing, and a small price to pay for having the rest of the year free on the Hill. While the stag shooting season ends on 21 October, contrary to some popular literature the rut certainly does not – and I have obtained some of the best observations and photos of rutting stags during the two or three weeks after this date.

By around 12 October the winds have usually switched to the first regular north-westerly gales and ravens have started flocking. These clever birds, which seem most stimulated in the north-westerlies, have learned over the years that the first weeks of October are when stags are stalked, and they will find plenty of large grallochs left on the Hill. Once when Moobli and I sat on a rock to eat lunch, a flock of twenty-three ravens came over the eastern ridges and flew very close. They looped and swooped, twisted their heads down and uttered muffled croaking noises, hovering on the wind to see if I was gralloching a beast, their great black shapes only about ten yards away. And I had run out of film. I threw out a few crusts, doubting they would bother with them, but as soon as we left they pounced down avidly.

To see wild stags mating in the Highland hills is a rare event and in the first ten years I only saw it three times, twice around dawn after overnight camps so that there was not enough light for photos. Hinds are in oestrus for only about twenty hours in a three-week period, but in this time they become very affectionate towards their lord and even seem to solicit his favour. The two stand close together first, she may nuzzle and lick his neck and muzzle, or even rub herself against his flanks like a cat against human legs. Greatly aroused by this he may lick her back, breathing heavily as if enjoying her scent. Then he moves towards the rear, pushes with his chin on top of her rump as he rears up, as if to give himself more of a lift, and gives a deep heaving thrust, maybe two, rearing straight up with the spasm, and then is down again and it's all over.

One 5 November, when an old grey stag and several hinds were using the west wood as a dormitory in cold rains, I was woken up not by fireworks but an oddly high bleating roar. I rushed out in the pre-dawn twilight to see the stag serving the hind and was sure the sound had been made by the hind, perhaps from a combination of ecstasy and pain. He slid off as she shot forward, shook his antlers and staggered on his right leg as if about to fall. Then they both walked away, he following five other hinds which were trailing unconcernedly up the gully to the northwest.

'Next time do it in the daylight!' I muttered gruffly as I went back to bed.

Although stag-fights during the rut are often mentioned in popular literature, I have not found them common in the hills round our home, for usually the roarings and menacing runs at rivals turn the lesser beasts away. When they do occur they are a riveting spectacle.

As intermittent hailstorms whipped spindrift from the loch surface one early November afternoon, I set off with Moobli through the steep wooded cliffs to the east towards where we had heard distant roarings at over 1,200 feet. Panting with exertion we rounded a ridge and peered through the heather and there, ahead, was the perfect fight scenario.

An ageing master stag, big-bodied but with only eight-point antlers, and holding only four hinds and two yearlings, was glaring upwards towards a huge boulder, making guttural '*ur ur*' growls. From behind the rock came loud blatting roars, higher pitched, then out stepped a younger stag, a fine ten-pointer in his prime at about nine years old. The old stag gave two loud roaring '*hrrumph*'s and moved to oust the intruder, which not only stood his ground but placed his forehead flat to the ground, swished his antlers to and fro, and threw up wads of damp herbage. Then he advanced resolutely. There was a pause, a slight turning away of heads as if both had thought better of it, then a swift turn in, and CLASH! The battle was on.

Stags don't jab and gouge like bulls but lock antlers, forehead to forehead, and strive to assert dominance by agility and strength. They heaved and strained against each other, rear feet whirling this side and that, eyes rolling, breath snorting, divots of earth churned up by the striving hooves. At first the older stag seemed to be winning, trying to lift the other and force him back on to his rear end, but the younger beast switched his back legs up the hill, the right leg finding added thrust against a tussock, and slowly the old boy's rear legs began to slip backwards, his front to buckle.

Twisting his head and antlers harder as if trying to weaken his adversary's neck too, the younger stag shoved harder and suddenly the old master's rear feet slithered into a muddy cleft. Feeling the insecurity of his footing, unable to match the thrust of the younger stag which now had the steep slope and the force of gravity to his advantage, the old stag gave up. He wheeled away suddenly and trotted off, the winner making a sweeping antler pass at his rear, as if to help him on his way, but missed.

Weakened by the long rut, the defeated stag, which would probably have overcome his rival at the start of it, turned at 300 yards and looked back. The winner now stood high on a precipice giving roaring vent to his triumph. The hinds were now his and he would have his chance with any that had not already been served by the master – or any others he could round up into his late harem.

Stags like to wallow in soft black peat during the summer and especially in the rutting period when they are not actually with hinds. I had seen this a few times but had never been able to photograph it. It wasn't until our seventh autumn at Wildernesse, on a rare and lucky 14 October, that I not only got wallow photos but also pictures of a savage fight in the open. We were heading back down the western side of the big dished corrie after stalking thirteen groups of stags and hinds when I saw a large stag amble over a ridge in front of us, his roars sounding muffled once he was over its far side. I was about to dash to the ridge before he got too far away when I saw a big black stag standing in a marshy area by a small lochan to the right of the ridge. He was roaring in the direction of the departed stag. Suddenly the first stag came back over the ridge but further to the left, driving a younger stag before him. The two sparred briefly before the younger one headed back to the south and both disappeared over the ridge again. Then the black beast set off southeastwards towards the first stag, which was clearly defending a harem of hinds, and also vanished behind the ridge. Now, thrilling moments these, we heard all three roaring.

Wondering what might happen when the two big stags met, I dashed dangerously with Moobli over the tussocks towards the ridge, but had barely covered 200 yards when the black stag and the ousted younger one both walked from behind the ridge and off westwards, towards the lochan. I ducked down instantly, wondering if both had been threatened away by the first stag. I waited until they were out of sight behind a small hillock a hundred yards to the southwest, hurried over to it while bent double, and took a few photos of the two stags walking along together in what seemed perfect amity. The younger stag halted while the black beast walked into a peat hag by the lochan, stood on the firm parts at the edge, lowered his antlers into the dark peaty sludge and, with his forehead also pressed down deep, switched them from side to side throwing up showers of the sloshy mush. As it dripped from his antlers he then ploughed at the sloppy liquid with his right forefoot, splashing it up against his rear parts. I now saw he was also urinating in a great spurt as if to make the mush even wetter.

All the while, as the camera clicked, the younger stag on the left stood roaring, not towards the stag in the bog but in the direction of the big beast behind the ridge with the hinds. Then the dark stag sank down into the thick black porridge, rolled from side to side and stretched his neck and dark mane out flat in it too, rubbing his chin with twisting movements on the edge of the bank as well. No wonder stags stink worse than old goats towards the end of the rut. Twice he got up and made hunching movements as if actually mating, trying to lower his parts into the mud. It was the only soft thing around perhaps, though I would have thought it somewhat cold!

After a few minutes he had had enough mud-bathing and wallowing and again walked off with the younger stag, back towards the ridge from where they'd come. Suddenly both stags halted, faced each other a few yards apart, the dark beast, awesomely blacker now with the wet peat daub on his coat, glaring at the other, as if suddenly resenting its presence. Then both went towards each other, turned heads away briefly, took a step backwards and with a simultaneous charge clashed their antlers together. It sounded like a pair of giants jousting with hat racks. Both withdrew slightly, as if setting their antlers right, clashed again, then began to push, hooves stomping and digging up the turf. I could hear their breath coming in snorts, see their eyes rolling wildly as I clicked away. They switched their rear legs about, each trying to find a slight uphill advantage or a tussock from which to gain added push.

Finally the younger stag's legs began to buckle and he slipped, recovered, slipped again, then turned to flee. The black stag gouged at his rump for good measure as he ran off to the south. The victor then walked towards the stag with the hinds and once more disappeared behind the ridge. Anxious to see what would happen this time – surely he wouldn't seek another fight – I dashed with Moobli to the ridge and looked over the edge. I was just in time for a picture of the black stag's head, so close I could get nothing more of him in frame, his eyes yellow, breath still spurting like a steam engine. He saw me then, as the winder scrabbled at the end of my film, and I was afraid he might attack. But he swerved wildly and hurried away.

What a memorable day, I thought, as we wended our way homewards, my knees creaking as I lowered my back-packed weight wearily into one sploshy hollow after another between the deep tussocks on the almost sheer slopes. I forgot right then that I had made nearly a thousand treks in ten years of Highland wilderness living, plus three years trekking after black and grizzly bear in Canada, to reach the stage where one could in one day approach closely and photograph fourteen separate stag-rutting groups in Britain's wildest mountains without them knowing one was there, not to mention recording a stag fight, and the climax of the day, a stalk so close all one could photograph was the stag's head.

By early November, the hills look tawny between the limp brown bracken patches, the flowing tussock grasses now yellow before becoming the sere 'white grass' of winter. Most of the big master stags have come to the end of their rut. They stand for long periods with drooping heads as if dazed and are more tolerant towards the younger stags. I have seen tired old beasts lying down, eyes half closed, taking no notice of four- or five-year-old stags which amble past, grazing less than twenty yards away. It's often believed the big stag is exhausted towards the end of the rut by the constant mating, but the truth is that it would be unusual for more than one, at most two, hinds to be ready for him on any particular day, and some of the younger hinds in the now dwindled harem may not yet have come into season at all. It is as much the constant running about, chivvying the hinds and bellowing challenges to rivals, and the lack of interest in feeding that causes exhaustion.

Now the big stags resume their feeding, their bellows limited to two or three desultory roars a day, often made when lying down. One by one they 'abdicate', grazing further away from the hinds until they amble back to where they meet other 'abdicated' stags and form their sociable wintering herds. It is not always a peaceful abdication, however, for sometimes the more mature of the young stags, impatient at having been kept away from the hind herds for the first few weeks, begin to move in and offer to fight to oust the older monarchs. Although the old stags are about ready to leave, they often won't tolerate such aggression from juniors. Once I saw a young stag with six points, which had been watching an old ten-pointer walking away from his hind herd, suddenly trot towards him with short bellows. The old beast turned, stood his ground with a look that indicated he was leaving of his own accord, not because of the other's threat, lowered his antlers and took two paces towards his young rival. The younger stag instantly wheeled and trotted back in the direction he had come. On the other hand, as already described, old stags can be ousted in brief skirmishes by younger stags a day or two before they would have gone by choice.

Now the younger stags have their chance with the hinds and there is a more peaceful atmosphere at the end of the rutting season. The roars are less frequent, higher and more bleating in pitch. On 11 November I came upon seventeen hinds with three young stags among them, all seeming quite amicable together. Fights are fewer and less violent and protracted, an air of amateurism pervades the young which come from these pickings, and I've seen mature hinds, now well served by the masters, peevishly strike out with their forefeet at young swains who show too close an interest. The amorous youngsters are filled with eagerness but their lack of experience shows, and they often look most disgruntled when their approaches are rebuffed. They have to wait until either a hind comes into second oestrus – if for some reason she was not served the first time – or for the younger hinds to come into their first oestrus.

These later pairings result in calves being born in late June or July after the usual eight months' gestation, but usually soon enough for the calves to be big enough to survive their first winter. An odd phenomenon is that some hinds which have not mated will come into oestrus at three-week periods through much of the winter, and may still find a stag to mate with them. Luckily these matings are rare, for the resulting late-born calves (up to late September) have little chance of surviving their first winter.

Normally by mid-November, as the first sleet begins to fall and the first gales of winter blast the leaves from the lowland woods, most of the hinds are in their own herds again, and within a few days there are a couple of dozen calves, yearlings and hinds spending the nights in our own little woods. An occasional juvenile roar or two can be heard from a young stag up in the daytime hills but before the end of the month nearly all stags' passions are spent and peace returns once more to the deer herds. It is then, however, they face the harsh trials of winter.

Chapter 19

The Mountains in Winter

The first hail showers, sounding as if a giant is pebble-dashing my iron roof, follow the teeming rains of late autumn. As the burns roar in torrents, showing up like thick white veins against the dark hills where one never suspected burns existed, the Highland trekker knows he faces six months of coping with the hail, engulfing mists and later, snow. One day a white mist advances up the loch, blotting the far shore from view, and soft rain falls for an hour followed by sleet. Then the air grows icy cold and wind flurries hurl down swathes of hail with huge stones bouncing from rocks and fallen logs, vanishing into the loch waters with a million plops and gurgles, hissing into and helping beat down the remaining vegetation on the mountain slopes.

Up high now we try to cover open ground between these violent showers, to walk with one behind and one before us and stay miraculously in the clear and always to look for the slight shelter of an overhanging rockface under which to dash if necessary. In November the first snows fall; we go outside and see the top thousand feet of the highest mountains whitening before our eyes, while the warmth rising from the loch and lowlands melts the snow so it falls where we stand as a soft drizzle and the 500-foot ridges behind the cottage remain as yet unscathed. On them, silhouetted against the dusky skyline, a long reptilian head topped with twin spikes and a few rounded forms betray the presence of hinds. They gaze down to ensure it is only us and there is no danger in coming low again to their dormitory in our woods.

A day or two of deep blue sky, the waning sun gleaming low and briefly over the hills before vanishing behind the far ridges, then the first frosts hit us. In early December the ground becomes frozen, the water-pipe freezes up and I have to break the glare ice in the burn or go down to the loch for my daily bucket or two. Now the fallen grasses crack beneath our feet, steam rises in the morning from the loch, warmer than the air above it, ice forms along its shores and icicles dangle from the rocks that line the gorge of the burn.

Trekking the high hills in these conditions soon reduces a man to size. Apart from the hail storms one has to contend with mist. One day we were traversing the highest ridges to the northwest when Moobli's high-stepping nose-high gait told me there were deer ahead. We rounded a small face and only fifty yards away were a huge hind and a calf eerily enveloped in fingers of mist. I backed down, stalked in a downward circle and had just taken three close pictures when the main mist stole silently upon our alpine world.

Suddenly the white hills beyond were cut off from sight and the mist became a shining white backcloth illumining first the witch's broom of a leafless rowan tree and an old stump, and then a strangely shaped rock which I would not otherwise have noticed. It was as if in its silvery glow the mist was revealing Highland treasures in a unique and personal one-man show. But such mist is treacherous, and all romantic notions fade when it thickens. I could see no more than three yards ahead. The light of the sky was uniform all round so that I was unable to detect where the sun was and, due to the circular stalking, I was no longer sure which direction I was facing. But for the snow that day we could have been in trouble, for the mist was to last all night. I did the only sensible thing and backtracked our own prints in the snow, heading down to the loch shore beside the first burn we came to.

When we were up on the high plateaus on another occasion, all the snow having been melted by warm sou'westerlies so that I couldn't use the back-tracking technique, we were again enveloped in thick mist. All wind had dropped, leaving us without any direction indicator. I headed over rough level ground, then down beside a big burn which I hoped was running south to our own loch. After a half mile I realised (from an old trick I had learned in Canada) we were probably heading *north* – for most of the tree branches were facing the upward slopes, therefore on the south sides of the trees. We headed back over the ridges again, trying to steer a straight line, until we hit a downward deer path that led to another burn. This time most branches were facing the downward slopes, and we followed the deer tracks beside the burn and thankfully came out on to the familiar loch shore. Next day I sent off for a mail order compass!

Trekking the high snowy hills is a totally different experience from any other time of year, and the wilderness dweller learns the necessary new techniques slowly. One early December morning, a dead white frost mist lay over the glens and corries of the hills, then a north-west wind sprang up, dark clouds scudded across the sky which became a steely blue-black, the mist cleared and within minutes a sleety snow began to fall. The wind dropped again and the snow fell straight down like flakes of zinc, lead heavy and covering the icy hard ground with a deep white blanket. Anxious not to have to put off a trek, I looked up into the hills, surprised to see they still looked their usual early winter tawny colour, with thin strips of white everywhere. Then as we plodded up to a 1,500-foot peak I realised why.

From below I had seen only the sheer sides of the innumerable tiny ledges and tufts of herbage; up on the tops the whole landscape was white, the snow covering every ledge or patch of grass, though it could only be seen as thin lines from below. Leaning against icy winds blowing on the tops, I had to fight for my footing and felt I could almost take flight. Below us a kestrel was hovering a full 300 feet from the ground and I marvelled again at its eyesight, and the necessary optimism that must fill the tiny hunter's heart – for what vole, or even beetle, would be out in such conditions?

Coming back from the peak, the wind now behind us, I learned how its direction could be changed by the snow. Peering over one ridge I saw a large hind right below me, so close I could have jumped on her back as she buried her muzzle in a tuft of grass. The wind was bouncing upwards from the flat snow-covered shelves before again curling down, and she was in a scentless pocket. She was too close for a picture, so I waved my hand airily and she took off upwards, stopping about forty yards away to take a second look. I got a fine picture now the snow had stopped, her grey velvety early winter coat nicely framed against the snowy hills and blue sky.

On another trek we reached the 600-foot level and nothing seemed to be in front of us but undulating plains of white. Suddenly some bumps on the ground ahead shot upwards, shook themselves in a misty blur of snow, and away to the west ran five hinds. They had been totally invisible. Having seen no tracks I had no idea they were there. Many times since I have seen small areas of slopes come alive in this way. The deer, having now grown their thick winter coats with their new linings of dense soft downy fur, can sit and chew the cud with snow falling until they are completely covered without apparently feeling the cold too much. The outer hairs get wet but when the deer are still in good condition, fat from summer and autumn grazing, the moisture never penetrates the downy layer. The snow even acts as further insulation against bitter winds.

As we climbed higher the clouds parted, a patch of blue sky appeared briefly, the sun beamed down until the bright light stunned the eyes, then all was grey and dark again. A wall of snow advanced diagonally over the hills towards us, a great grey curtain, and suddenly we were stumbling for an overhanging rock face. A single red grouse shot up from an igloo formed over some heather and with protesting clucks weaved low between the ridges and vanished. I sheltered below the face, my body heat from the trek soon sucked out of me by the wind, while Moobli just stood out in the snow which settled on his head and eyebrows, giving him an ancient mariner look, as he blinked at me wondering why I had stopped walking.

'You know something, Moobli,' I said through chattering teeth. 'One is no nearer heaven by climbing a mountain.'

Stalking wildlife in snowy conditions demands different techniques. I soon learned that frosty snow, hardened on top overnight, crunched loudly underfoot and deer could feel the ground vibration; when the wind was low they could even hear it a quarter of a mile away. It was difficult to avoid walking on snow for it covered all the tiny ledges which comprised most of the steeper slopes. Although irregular they provided little steps for climbing, and even when there was little or no snow they were still filled with ice. Always I sought grassy tufts projecting beyond the ice and snow to give extra purchase for my feet. Often, being frozen too, they snapped off, so I used gloved hands as well, my weight distributed on two or three points at a time rather than just one foot. Whenever the sun came out I learned to try to trek along the south-facing slopes. Towards the end of the afternoon we could be walking in soft mud and wet grass, like on a damp autumn day, but on the north-facing slopes which had not felt the sun the tussock grasses were as thick with grey hoar frost or snow as before the dawn, all the little ledges still filled with treacherous ice.

For weeks at a time the high hill lochans were totally frozen over with ice so thick one could have driven a bulldozer across them. I learned to follow deer tracks on steep inclines as much as I could, not only because deer seemed able to sense the exact depth and the firmness of the ground below but because I could *see* exactly how far down it was and prevent the snow going over the top of my boots. I tried the usual leather mountain boots and gaiters but over ten miles or more the boots, no matter how well greased, became soaked and freezing cold to the feet, and the gaiters trapped snow too. Good old rubber wellies became my stand-by and I wore out eleven pairs of them in the first ten years, matched by six pairs of hill boots, which I used mostly on summer eagle treks.

When traversing the steeper sides of the hills in frosty snow I found I could make my own secure steps with a light stamp of each foot at any angle I wished, and had no need to worry about the contours of the tussocks, now safely beneath the smooth blanket. But after clear cold nights this snow became so hard-packed and frozen that the hardest stamp barely left a mark. Then it was impossible to cross these patches without an ice axe to hold on to, or I could be sliding down to the bottom – as indeed I did at first. Some slopes ended above precipitous drops and I had to labour round them rather than take dangerous chances.

Crossing the deeper ravines on the frozen high tops was also hazardous. Spraying water from the burns covered every rock with ice, and often the burns had frozen up too. One false step and one would be away to a slithering painful death on an icy staircase with rocks projecting like teeth all the way down.

Yet to be up high in the mountains in the grip of winter is to know a magically different world. As we set off one sunny January morning, the small cliffs of the burn gorge were festooned with icicles, many hanging from grassy ledges and loosely from the herbage itself, so that in the breeze they moved and struck each other with a beautiful chiming music. Higher up we came to a small overhanging cliff where huge icicles projected like fangs, turning rock faces into enormous gargoyles. Like huge crystal organ pipes over six feet long, the icicles dwarfed Moobli as he walked between them. I paused to look back: wisps of steam were coming up from the loch and drifting along above the surface like carpets of silver. The sun shimmered with tiny stabs of light upon the water furrows, round the islets, beating up into the eyes with intense fractured light. Small waterfalls rustled faintly under the glare ice of their burns in the distance, and above us the sky held the impenetrable blue of eternity. With a sharp chinking cry a lone ring ousel flew past us to the east, pointed wings and diamond tail trailing, while on the broader ledges we found the deep slots of deer-made trails in the snow, unsuspected from below.

Higher up, the icicle displays from the craggy faces were even more spectacular, glinting like rainbows, dripping water from their pointed tips into tiny snow pools. A thin silent sound as of harps seemed to pervade the air. Now and again an icicle snapped off and tumbled down the snow-filled slopes to startle the lonely wanderers.

A crisp inspiring day and I was feeling good, so we hiked three rough miles until, just south of an 1,800-foot peak, we saw deer on the far west-facing slopes across the deep river valley and I decided to stalk them as the sun would be well behind us by the time we got near. The wind too was ideally from the north-east. Not until we reached the fenced forestry compartment did I cheek with the fieldglass again: beyond the herd I had first seen, all hinds and calves, were three stags. This was a break for they were two miles nearer us than their usual wintering corrie.

We cut into the forestry section, Moobli squeezing his massive bulk sideways through a six-inch gap above a drain, then crept in zig-zags along the drains. South-east down one drain for 300 yards, then north-east up another for 600 yards towards the deer. We repeated this several times until finally we had enough sizeable spruces to screen us to get to the far fence unseen. Then we had to creep along the fence until we could put a small ridge between us and the deer. I climbed out below their sight line and Moobli again did his scraping squeeze along a drain.

We cut below more small ridges, then up on to a grassy hill. That left the hind herd between us and the stags and it was impossible to get near the big beasts without the hinds giving the alarm. Back down to the west I had to go again, along another drain and fence, crawling on hands and knees. Hell, it was cold, my knees frozen, my hands blue even though I was crawling on my fists in poor imitation of a gorilla, to save my finger tips. We went up the edge of a frozen burn to the east and on to another grassy hill. I now took four 'insurance' photos of the nearest stag, an old beast sitting down. Obligingly he got up and delicately scratched his ear with a large rear hoof. Then two of the stags started moving about, as if sensing something was amiss. Yet again we had to cut down to the west, into another forestry section and over its far fence, where I saw a huge square rock all alone on the hill at 1,500 feet. It was hands and knees again until we got up to it.

The sun was now blinding the stags, eighteen of them straight ahead, all lit up like lamps against the patchy snow. There were nine more stags a little further ahead – twenty-seven in all, the most I had so far seen together in the winter wilds. Blinded by the sun they could not have seen us if we had stood up. I got pictures of them grazing, shaking their coats, scratching themselves, or just basking in the sun. I could choose between groups of nine, seven, four or just pairs – for the first time I was actually able to worry about the sheer *composition* of wild deer pictures! No wonder I had not known there were so many for even at a fairly close distance their red-browny coats blended so perfectly with the glowing wet sunlit bracken clumps on the hill. I took about thirty lovely photos and left in the ideal way, without them ever knowing we had been there. A memorable stalking day.

Snow's finest quality for the wildlife-watcher is its ability to preserve for a time the tracks of animals, revealing their nightly movements and resting places: where deer stopped to feed, where a badger scraped through leaf litter and under rotting twigs to get at worms or chrysalises, where a fox pounced on a vole or beetle, took a side step to leave scent on a rock, or where, more rarely, the wings of a crow or raven hit into the snow and left beautiful wide brush marks as it landed on some small prey. Yet, apart from deer, it is useless to try to snow-track mammals like fox, badger or wildcat after the first fall for they seldom venture out the day following, as if hoping after a prolonged 24-hour sleep the snow will be gone next day. When the snow persists hunger forces them out.

I learned that wildcats and the rare pine marten will always use a fallen tree as a bridge to cross burns if they can, rather than jump from rock to snowy rock, that a fox will try to avoid crossing snow, wending its way between the patches or around white meadows by skulking under trees along burn banks whose shelter helps break up the falls. From snow-tracking, too, I discovered a fox will circle round a fresh deer carcass for three nights in succession but not actually start tearing into it until usually the fourth, as if fearing a trap. In open areas in the woods we often found snow tracks of blackbirds, robins, chaffinches and tits, along the sides of fallen logs where they had been probing the undersides for insects and beetles.

On one high snow trek I was perplexed to find tracks about the size made by small deer along narrow ledges, but they had melted slightly, snow had fallen into them and so they were obscured. Between the tracks were long scrapes as if made by a heavy tail. I recalled reading somewhere that otters will cross mountain ranges, even in snow, and their tails left marks like these. Mystified, I followed them until I saw a small red deer hind and her calf ahead. As I watched them walking, I saw their rear legs swept inwards slightly with each step. They were too tired to raise their feet to clear the snow and the deep scrapes had been made by the incurving action of the rear feet. For me, that ended the otter going over snowy mountains theory. As we headed to a higher ridge, three redwings hopped over the snow in front of us, their wings trailing down as if too heavy to hold up, before they finally flew downhill. They and meadow pipits had been eating seeds from the dry brown heather shoots that stuck up here and there above the level of the crisp snow, the faded husks scattered around each shoot.

One unusual experience was to find deep single tracks over a foot apart that led in a straight line between rushes along the small burn that edged the west wood. At first I thought of the long-legged fox but the deep holes were too thin, and as snow had fallen into the tracks it was impossible to make out the prints below. In this way a fresh track can sometimes appear old. I kept watch early next morning and saw a heron stalking along. When it flew away I found the same kind of tracks where it had been walking.

Sneaking round the woods with Moobli early one January morning, I saw a movement by a deer calf carcass that lay in the open across the main burn. The large female buzzard was standing near it and she had not spotted us. As we watched she jumped heavily on to the carcass, then swept the snow off backwards with her feet. She set to work with her hooked beak, wrenching off morsels and jerking them back down her throat. When she had gone I hastily erected a hide in the 'tents' of broken brown bracken and a few hours later, after making sure the buzzard was nowhere about, I slid into it. Usually when working a hide for such a sharp-eyed bird one needs two friends to put one in. The bird sees humans walking away and then thinks the coast is clear. Alone I could not do this. I endured three bitter hours in the hide but not once did she, or anything else, put in an appearance. Not until I went into the hide in the pre-dawn twilight did I have any luck.

After almost an hour on the first morning I heard a slight sound of rushing air, and saw the buzzard land in an oak tree on the far side of the clearing. She stayed there immobile and almost invisible for a further infuriating half hour as my extremities became colder and colder. She seemed as cunning as a hoody, making sure there was no danger about.

At last she launched herself and glided down to land a good seven yards from the calf. She then walked round it on her thick pantalooned legs with slow deliberate steps as if to make sure it wasn't booby-trapped! Finally she stood on the stomach and some entrails which a fox had removed two yards away and began to rend the flesh with strong upward heaves of her beak and neck. Each morsel she swallowed by holding her head almost straight up. She ate for eleven minutes, constantly pausing to look around and listen for danger, then with ponderous wing-beats flew away.

When I went over a quarter of an hour later I found she had been eating the entrails and some of the stomach skin, which seemed odd with so much fresh red meat still on the actual carcass. Although I had taken several pictures, the needle of the light meter barely quivered at an eighth of a second, so the pictures wouldn't be any good.

Shortly before dusk on the next afternoon I saw a buzzard I thought was her fly past the window above the lochside trees. I hurried out and watched her beat her way right over the carcass and disappear over the far trees to the east. Grabbing the camera, I dashed over the slippery rocks of the burn and into the hide, hoping she might come back. It was an unusually lucky decision.

Just as dusk was falling at 4.45 p.m. the buzzard came back, landed in a different tree, then after only a brief look round flew down to within a foot of the carcass itself. The bird took two determined steps, grabbed the flesh hard with one foot then stamped down with the other, bent down with comical slowness, took a bite, then hauled upwards. When the meat came away suddenly the bird overbalanced and had to flap her wings. I saw then it was a different bird, its wings far whiter underneath, and I was sure it was the chick I had watched at the nest the previous summer, also a female for she was as big as her mother. It was clear she was not yet experienced at rending carrion.

She too looked about her after every two or three beakfuls, swallowing with jerky forward movements of her head or tipping up her beak. She fed for a good twenty minutes while I raged inwardly at the awful light. Once she walked away a few paces, stiffly like a parrot, as if she had eaten enough. Then she looked back at the meat again, evidently thought she would have some more, hopped gawkily, almost ponderously, on to the ribcage and tugged off a few more bits. I thought what a beautiful creature

she was, the incarnation of wildness, her great brown eyes looking black in the poor light, tugging so hard I could almost feel the
vibrations. It was amazing how much meat she could put away,
and I felt she knew there would not be much left after one more
night with the foxes at the carcass. Unfortunately my camera
needle still only registered a fairly useless 1/30th of a second but
I took a dozen photos. I was surprised at the power in her neck
and back muscles, and was even more so after she flew away and
I found it was all I could do to rend off meat fibres in the same
way with my fingers.

I only used the hide once more, defeated by the cold, but nothing came down, nor did I see any buzzard, crow or raven go near
that carcass in full daylight. Although the east wood lay between
it and the cottage, it seemed it was still too near a human dwelling for the birds to abandon caution.

I once witnessed at close quarters a buzzard actually killing
prey and it was unforgettable. I was driving down a quiet single-
track road flanked by open woodland and had just rounded a
bend when I was astonished to see a buzzard drift across in front
of me. With its wings half closed it really floated, very slowly, as
if in slow motion, then it hit into a tussock clump on the left side
of the road. It struck twice rapidly with one foot, looking fierce,
its head up and beak down, and I saw the head and part of the
dark zig-zag pattern of an adder as the snake lashed about trying
to strike back with its fangs at the big bird.

Oddly, the buzzard moved, almost danced, with seemingly
exaggerated slowness, like a matador out-thinking a bull, then
struck with lightning speed at precisely the right moment so as
to sink the talons of its other foot and get a shorter grip on the
adder's neck. The bird was not going to give up, despite the
Land Rover's closeness. I was past in seconds and knew better
than to stop suddenly and leap out with the camera and telephoto lens, which for once were on the seat beside me. Instead,
I stopped gently, let down the far window and seized the camera
– but I was too late for the buzzard was away through the trees

holding the writhing snake just behind its head with one foot. What most impressed me was the way it clutched its talons together hard after each down strike, showing it killed with them like the eagle. It also displayed a strangely detached air, as if not lusting to kill but simply fulfilling what it had evolved to do naturally – snake and hawk linked together in a brief dance of death.

Perhaps the most unusual tracks we found in the winter hills came on the snowy afternoon when for two days running I had seen a young eagle fly weakly from the west and land on the highest ridge of the skyline to the north. Hinds and calves had worked their way up to those ridges after spending the nights in our woods, and the eagle seemed to be using the peak as a spy post. It watched the deer, its head following their movements, as if hoping a calf would drop. When the eagle landed there again on the third day, this time with a few ravens and crows apparently following, I decided to stalk it for a photo.

We moved through the west wood, a nearer ridge in front of the eagle now becoming the skyline, and headed slowly upwards. The wind was right too, so we didn't displace any of the deer. When we emerged carefully from a deep rock fissure near the top we were just too late – the eagle was winging away westwards and some ravens and hoodies flew up from the nearby grass clumps and heather tufts. We walked over and there was a deer gralloch left by the estate stalkers who were now culling the hinds. But the bonus was the discovery of deep eagle tracks in the snow, each a full six inches in length, showing where the eagle had landed, had tramped west to east, then to the south, then back north-west, crossing its original track towards the easy feast.

Most of the stomach skin had been eaten away from the gralloch, and only the half-digested grassy mush was left on the surface. It was interesting that the eagle tracks were on their own. Evidently the eagle had beaten the other birds to the gralloch. Although I had read stories of ravens and crows mobbing an eagle from its prey, it appeared they had left this hungry eagle well alone. Their tracks among the grasses and heather were four yards from the gralloch, as if they had kept at a discreet distance until the eagle had eaten its fill.

Chapter 20

Winter of the Deer

When one lives for years in the remote Highland wilds without most modern conveniences, can only reach home by a long boat journey often in biting cold, hail, rain or gales, has to cut wood to keep warm, and treks the winter hills, one is close to the animal state oneself. It is impossible not to feel a deep sympathy for all the creatures of the wild, but perhaps over the years I felt mostly for the herds of graceful deer.

Red deer were originally woodland animals but centuries of burning forests to drive out wolves and rebels, clearing space for agriculture, roads and homes, felling trees for charcoal, kilns, and ships and building timbers, decimated most of the original Highland forests. The deer have been driven to adapt to existence on the open moors and rocky hills which, apart from some cliff faces and ledges, contain very little cover; yet archaic jargon persists for these bare hills are still called 'deer forests'.

Scottish red deer *have* adapted but are among the smallest in the world. It may seem a matter for self-congratulation among those concerned with land use that the deer multiplied from about 155,000 in 1960 to some 270,000 by 1978, but a succession of mild winters since the late 1960s had as much to do with it as creative management. In the exceptionally harsh winter of 1978/79, for instance, there was a rapid drop and it was estimated 15,000 animals died then, bringing the total population (at time of writing) to some 255,000.

While increasingly new forests are being planted, mainly by the Forestry Commission and large private estates (which obtain tax relief to do so), most of them of close-stacked conifers, deer are fenced out of them, and these artificial forests occupy the very lower glens which deer need for shelter and food in the harshest weather. Thus the deer have been deprived of even more vital land. In my own area thousands of acres have been removed from deer use in this way over the last ten years.

Red deer are prolific. On the open hills a hind has her first calf at about three years old, then calves only every alternate year, so about thirty-five calves are produced annually to every hundred hinds. Hinds with access to forests increase in size, can calve at two years old, have calves every year; the production rate then goes up to about sixty-five calves for every hundred hinds. Then the culling rate goes up from a one-sixth cull to a one-fifth cull, to maintain herd size.

Because they are prolific and have no natural predators left, culling by man is not only of high economic importance to the Highlands, but vital for the good of the herds themselves. Scotland's red deer occupy some 250 private estates and other lands and range over 2,500,000 hectares. About 12,000 stags are culled annually as well as between 14,000 and 18,000 hinds, including smaller numbers of calves, yielding about a thousand metric tonnes of dressed carcasses. The revenue from the venison, supplemented by the sporting fees of stag hunters, adds up to a substantial sum.

Nature's method of controlling numbers, and maintaining genetic fitness, is to eliminate the old, the weaker young and the sick, so allowing the fittest to carry on the species. The best modern stalker (usually trained by and following the advice of the largely excellent small-staffed Red Deer Commission) operates on this principle and tries to get to know his deer as well as possible. This is not easy when, faced with rising costs and recession, modern estates can no longer always employ a full-time stalker, and the man may have to double as a fishing ghillie, general keeper, fence and boat repairer and general handyman. Stalkers also change jobs more often than in the past. It takes

several years to know a Hill, and in the gaps part-timers sometimes have to be employed who have to take whatever stag or
hind they can get near to in a given time. The best stalker stoutly
resists the wishes of his shooting 'guests' (now increasingly
higher fee-payers from the Continent) to take only the best and
biggest beasts, and he ensures there is always a fair excess of
hinds over stags. He will cull the poorer quality animals of both
young and old age groups, concentrating slightly more on the
older hinds and poor younger stags, including the worst 'knob-
bers' which probably won't grow into good adult stags anyway.
In the past, when sport came first and venison was merely a
by-product, only the biggest and best antlered stags were taken.
This constant removal of the genetic strain of the big fellows over
many years has also had considerable effect. Yet weight and
antler size is as much due to nutrition as genetic inheritance.

Sometimes, when I have advocated supplementary feeding of
deer in hard winters, I have been accused of wanting to maintain
artificial stocks which could not survive naturally and which
would cease to forage in a normal healthy way. But there must be
a difference between creating artificial herds (the hinds of which
can be more easily and cheaply culled at the feeding areas during
their winter shooting season anyway) and a little supplementary
feeding of hay, mineral licks and root crops, to help deer over the
harshest period. The best red deer in the Highlands are certainly
on the estates where grazing competition is kept low, forest cover
is available in winter, and some supplementary feeding is made.
After all, man cushioned by centrally heated homes, roads, electricity, gas and transport services, lives in artificial herds too – yet
is still in need of supplementary benefits.

Apart from a little winter feed, I am convinced forestry owners
should allow deer into special small pockets of maturer woods
for winter shelter once the trees are over twenty feet high, and
after the first pulp-wood fellings, as well as plant or leave some
'weed' trees and bushes for their browsing needs and rubbing
posts. At present, when fence posts begin to decay after ten or

twelve years, hungry deer break in and inflict great damage over a large area. They are then hard to locate and control. Once I found an eleven-foot section of the fence round the plantations in the big river valley broken down by the stag herds, some of the spruces de-barked, tree-tops chewed and many branches stripped of needles as if a vacuum cleaner had sucked them off. The scene could have been called 'The Stags' Revenge'. A little supplementary feeding, opening an acre or two of forest for shelter, might have prevented this devastation. It would certainly have improved the quality of the deer in the long run.

Oddly enough, red deer do not suffer as much from crisp, snowy, freezing winters as they do from long, cold and wet ones, when the rains persist well into spring. The constant wet on the ground and in their coats, causing loss of heat through evaporation in the winds, is a far worse affliction than dry cold.

Should the weaker animals last through January and February a new problem awaits them for trees and bushes are leafless and the grazing herbage, often covered by snow, is at its poorest. It is a sad sight to see the weaker deer standing forlornly alone, scraping fitfully with a foreleg to try to reveal some food beneath. It is at this time that the greatest mortality occurs. Some animals, weakened further by combinations of liver flukes, infestations of warble fly larvae which make holes in the skin, nasal bot fly grubs which block the breathing passages, and lung worms, become unable to follow the herds on their daily upward migrations to heather slopes. They die on the open ground, often tucked between large tussocks, or try to seek shelter under any tree or rocky ridge they can find and finally die from a combination of starvation and pneumonia. In late April and May, when the new grasses start to grow and bushes and lower tree branches sprout their first leaves, a new phenomenon occurs. The still weak animals which have survived thus far naturally graze and browse greedily but their digestive systems, unused to the richer food, cannot cope and the deer become gut 'blown', and this also causes death.

Understanding these problems, I tried to help the deer a little. Apart from leaving the woods unfenced and caging my new trees, I also cut the summer hay in my front pasture with a hand scythe, stacked it on a three-pole pyramid, and in winter carried it to the woods, along with sheep cakes. As a result estate stalkers, cruising in a boat with binoculars to spot hinds, saw the bigger groups round Wildernesse and moved in to do their culling. I didn't think much of this; not only did the sudden bangs of the gunshots disturb my writing periods and scare away other wildlife I was studying, but I hadn't taken the trouble to help feed the deer to have people come along and bump them off! But they were nice enough men, doing a fairly hard job in winter, and I naturally kept relations cordial. All the same I learned only to feed the deer when the winds were unfavourable for the stalkers.

While every winter I found several dead or dying deer in my area, mostly old hinds and late-born calves, there was one exceptionally wet and tragic year when the plight of red deer at this time was brought home to me in unforgettably forceful terms.

On 22 January, after two weeks away on a business and research trip, Moobli and I came back to find a newly dead yeld hind by a fallen rowan in the east wood. Holes made by shotgun pellets in her right haunch and lower abdomen seemed to indicate she had been shot by an amateur poacher and had stumbled a long way to our woods before succumbing. On our walk round we found that the cages round one of the graceful hemlocks and two of the Douglas firs had been kicked down, the tops had been bitten off and the branches stripped of needles by deer's teeth. My initial anger at the ungrateful deer lessened as I realised I should have made the cages stronger. We also lost a snag-born silver fir I had replanted south-west of the cottage to give us future wind shelter, fondly believing its five-foot tip was beyond a deer's reach. That was before I knew red deer can balance like ballet dancers on their hind feet while nipping off twigs and the leader shoots of trees!

The first hint of impending tragedy among the deer came next morning, when four inches of overnight snow blanketed the ground. We were tracking the movements of our wildcats in the west wood, noting there were now fifteen hinds and calves on the slopes above, when Moobli lifted his nose, scenting something ahead in the north-west breeze. Keeping him back, I walked on and found a deer calf lying in a slight hollow. It was thin and cold to the touch and was seemingly at its last gasp. While it could just lift its head and make slight flailing movements with its upper foreleg it could not get to its feet. It was heavy and wet but I managed to get it across my shoulders in a fireman's lift and carry it over the tussocks through the wood and up the path into the cottage. It drank a little warmed lamb milk through the feeder bottle I kept for such emergencies but it was dead within an hour. Disliking waste, I dragged it to the burn, bled and skinned it, and removed the legs and haunches for Moobli, myself and the wildcats. As I tacked the hide over a pole frame to rub down with alum and saltpeter, I found no fewer than fifty-two warble fly pupae penetrating the skin – nasty little oval grubs encased in whitish yellow sacs of jelly, which must have also weakened the late-born calf. The skin wasn't much use but would make a footrug. I towed the rest of the carcass up into the open at 300 feet as a gift for the golden eagle pair I was studying.

Our next extraordinary experience came only five days later. We were walking up to check the carcass when I suddenly saw some hinds and calves running east towards the burn high above us. A small calf at the rear swung downwards, fell over in a tangle of legs and snow as if it had tripped on a rock, got up again and ran on down a cleft after the others and was lost to sight in the trees. As we hiked up, the large female buzzard flapped away from an oak above the burn. She had been watching the carcass but nothing had yet touched it. Suddenly Moobli gained an odd scent and began to run downstream. As I followed along the top of the gorge I saw the calf in the middle of the raging burn. It must have lost its footing following the others and fallen in, and but for Moobli's scenting powers I would never have known it was passing below.

It was swept over two fairly gentle falls into deep surging pools, its head feebly bobbing under twice, unable to get out. Realising it was approaching the big 25-foot falls below our water pipe and that it would be killed if it went over, I took ludicrous chances, chased down, dashed waist deep across the burn and down the easier far side. The calf fell into another swirling pool but managed to get its front feet down on the edge gravel and just stood there, miserably bedraggled and certainly weak. I scrambled over slippery rocks, clutched it by the neck and edged it across the burn at the widest point. Then I heaved it out, and with one hand on its rump and one on its neck managed to walk it a few yards before it fell down again. It lay there, exhausted after fighting the cold waters, its eyes rolling and showing much white at their rears, as much from shock as weakness.

I picked it up in my arms, draped its heavy wet body over my shoulders and, as I puffed down to the cottage, was deafened by a loud `bleeea!' in my ear, like a lamb bleat but far louder. In the kitchen I dried it out by the stove, needing two towels to soak up the wet, then warmed some lamb milk in the feeder bottle and, holding its head gently, put the teat into its mouth. To my delight, when it felt and tasted the warm fluid it sucked eagerly and soon swallowed nearly half a pint, its stomach rumbling as if it were near starving. Then it kindly urinated over the floor, from its middle, which made further examination to confirm it was a stag calf unnecessary.

It had no sign of antlers yet and was clearly a late-born calf. I wondered if its mother had been shot. After I gave it some more milk, it snuggled close, nuzzled my beard, laid its head on my arm, sighed, and closed its long-lashed eyes. A dear little thing. I put thick hay in the warmed up workshop, went to the garden to gather best green hay, some kale, sprout and cabbage leaves and two handfuls of sheep cakes and took them back to the workshop. To my surprise, as I arranged all this food on the floor the calf moved forward and with lightning speed, as fast as a wild-cat, smacked my hand with its right front hoof. It was a downward blow made with the flat of its cloven hoof so that it didn't hurt, though it just missed my face. It was strange for it didn't try to butt with its head or bite; just that swift slap, then it looked sweet and gentle again. It seemed the strike was simply an instinctive reaction to the closeness of a human, even though it now recognised I would not do it any harm. I kept up the soothing talk and during the day went back a few times to give it more milk, into which I put a little aureomycin powder to combat any viral infection. I was careful not to bend down too suddenly near it in case it hit out at my face, for it moved so fast, and without warning, one couldn't know it was coming.

I left a candle glowing dimly on the workshop bench and when I returned at 8 p.m. it had eaten a good deal of the vegetables and sheep nuts and hay. I gave it more milk and withdrew to cook my own meal.

Suddenly Moobli gave a low whine from his bed. The staggie had trotted out of the workshop, hooves clicking loudly on the concrete floor, and came to stand close by me as I peeled potatoes at the sink. Moobli moved forward but with a gentle expression on his face and sniffed the staggie's muzzle – and the calf seemed not the least frightened but sniffed his back! It seemed amazing it should accept so quickly and without panic the kitchen, myself, Moobli, and all the strange objects it had never seen before. I hoped it was not just due to weakness, though already it seemed stronger on its feet. It appeared not to notice the few flash photos I took before giving it more milk and walking it back to its hay bed in the workshop.

Next morning, the weather dull but milder, the calf was even more spry and snorted through its nose when I went in, but the tameness of the night before proved an illusion, for it again swiped out with a forefoot as I went near. I fed it more milk, then put one of Moobli's giant dog collars round its neck with the idea of letting it graze in the garden all day under supervision. I tied it with a long rope to the southern rhododendron bush. At first it bucked about, trying to haul itself backwards through the collar but, after tiring itself again, it settled down and grazed peacefully. When I looked up from my desk it was often sitting down chewing the cud. At night, with the sky clearing and a frosty nip in the air, I brought it back indoors. Now it only took two swallows of milk, ejected the teat from the side of its long thin jaws and refused to have any more. I left it in the warmed workshop with more garden greens.

Next morning I saw the small herd of hinds and calves back on the upper ridges. I walked the staggie out and he saw them too and tugged at his collar. I let him go and up the north hill he went, head bobbing up and down until he was trotting among the herd. One of the hinds shoved him away with a head bunt, and another struck out waspishly with a forefoot, but a third accepted him after a few suspicious sniffs. Then she allowed him to nuzzle at her teats which still held milk.

Next day dawned in a clear sky, our water supply was frozen up again, and I saw the herd now grazing just above the west wood, bathed in brilliant light. We stalked them through the wood's dark shade and got to within forty yards for some really close-up pictures. And, yes, our calf was there, smaller than the other calves. He had been accepted back and seemed in good shape again.

On 6 February I saw two crows wheel down from the sky and disappear into the edge of the long wood along the shore to the west. I hiked over later and found an old dead hind, blood trickling from her nose. As there was no sign of shot or injury, she could have died from pneumonia induced by an infestation of lung worm. Six days later, in the north-west sleet squalls, I found a dead hind calf floating in the pool which fed our water pipe and towed her amid the bracken higher up for eagles or buzzards.

Boating up the loch in misty sunshine from a supply trip next day I saw something moving in the water nearly a mile ahead. Probably just a floating log, I thought as I stopped to put Moobli off for a short 1½-mile exercise run. When I started off again I saw the object had moved nearer to the north shore. It appeared to be humping oddly with up and down movements. My heart skipped a beat for an ancient legend held that our loch housed a monster and there, in the mist, it seemed I might just be seeing it! As the boat surged on the creature seemed to be on its side, like a big pike with an air bladder too distended to dive down, or which had been wounded. As I made ready to sheer off to one side and gun the engine, the object seemed smaller than it had looked from a distance and I saw now it was a deer hind. It had been trying to swim the loch but couldn't make it in the icy water, and it was drowning before my eyes. Even though I had slowed down, the boat was still travelling too fast and I had to reverse it back to her. She was still twitching, eyes distended, and when I grabbed her head to lift her nostrils clear of the surface, water billowed out of them. I towed her to shore and hauled her out. I had arrived too late and she was dead. I took the carcass home, skinned it, took the haunches for the animals and was towing the remains to an open area well away from the house when I found another dead stag calf that had been washed over the jam of fallen trees at the start of the secondary burn. Its body surged up and down in the boiling water.

Two days later we trekked over the high peaks to the river valley to the east and were stalking some hinds in a sunny spell when a government survey helicopter buzzed over with its familiar loud chopping sound. To my surprise the stalk wasn't spoiled for the deer took almost no notice, just a few calves darting about for a moment. They had evidently come to realise no harm came from these noisy machines. On the way back along the loch shore we found more tragedy. We had just passed the wreckage of a boatshed, which appeared to have been blown from its foundations beside a natural inlet, when we found the remains of a deer calf, all bundled up like a tangle of rags. Forty yards further on Moobli sniffed out the carcass of a big old hind well torn apart by foxes – the skull lying fifty yards to the north, the bottom jaws three yards to the south, while the lower parts of all four legs lay scattered ten yards east of the jaws. Each leg had been bitten off at the joint, the foxes cleverly cutting through the tendons rather than having to chew through the bones themselves. Along the path in the woodlands a newer carcass, of a hind calf with a yellowish bile issuing from her nose, lay in a rock cleft ten yards east of a badgers' sett. In a break in the trees, lying in an open hilly spot between tussocks, was yet another deer calf corpse, also well chewed by foxes.

To find a freshly dead or dying deer in winter was normal enough when living in the Highland wilds, and could even seem like a bonus when, with a full vegetable garden, bottled wild fruits, nuts and dried fungi, the extra meat made it almost possible for Moobli, the wildcats and myself to live entirely off the land. But already, with winter only half over, the death toll of deer was more than twice what it had been any year before, and I began to feel some alarm.

On 25 February I woke at dawn when I heard Moobli whine from his bed in the hall. I looked out of the window, just in time to see a deer hind, her head down, stomach swollen, wandering close to the cottage towards the east wood. I hauled on some clothes and sneaked out to the front door. She looked dead beat, walking slowly, brokenly, as if about to collapse. Shutting Moobli in, I stalked her in wind and drizzle and was just in time to see her walk into the rhododendron bush north of the spruce glade, collapse on her front knees, stay there a moment then fall on her side. I hurried back for my camera and fitted on the flash gun because of the poor light, but the batteries were so low the charged light would not go on. I went out again and crept towards her, feeling I could catch her by hand when she was so fatigued and try nurse and feed her back to health.

She was now in the death throes, her neck arched back at more than right angles, feet kicking out. My touching her would probably only make things worse, so I kept back. After a few seconds I couldn't stand just watching and went over to her. As I cradled her neck and head, she gave two spasmodic kicks with her rear legs and her head dropped earthwards, dying in my arms. She was very thin but I could see nothing else wrong with her externally. As I looked down at her lifeless form I thought how tragic it was that when they needed shelter, or knew they were about to die, the deer came to our little woods. Being originally woodland animals, when death was near back they came to their natural environment. She had died deep in a rhododendron bush, surrounded by swaying vegetation, her eyes seeing the last of life and the sky through the soft green sheltering leaves. And just behind the bush, nodding in the wind, I saw the yellow head of the year's first daffodil. Sad, I didn't feel like skinning yet another deer, so I hauled her up the slopes for the eagles.

Nothing more happened until 11 March. Then on our early morning walk I saw a hind swimming in the loch, way out past the gulls' islet, as if she had swum all the way from the opposite shore. She bravely fought her way to the beach about a quarter of a mile east of us and I thought she'd walk out and trot away. But she stumbled on the first rocks with only her back clear of the water, went under, got up again, tried to go further but stumbled again, then flopped down with some splashing. This time she didn't get up. What was wrong? I went back to shut Moobli in the cottage and put on my raingear as it had begun to pour heavily. Then I forged through the swollen burn.

I found the hind lodged on some large sharp rocks. How cruel they must have looked to her after her long icy swim. She too was in the last throes of death, mouth opening and neck stretching but her whole head under water, her eyes black pupilled, wide open, and bulging with terror and incomprehension. I dashed down and tried to pump water out of her lungs after dragging her out – but to no avail. Another goner. Again, I was too sick at heart to skin her or even take her haunches.

The very next day we were returning from a search for dens in the woods east of us when a small hind yearling got up from the old bracken in our own wood and tried weakly to run. I sent

Moobli to turn her, which he did in a few seconds. She went down on her knees and I caught her, then with Moobli keeping well back – for he now knew deer would refuse to budge if he was in front of them – I walked her up to the cottage. By now I had decided not to put deer in the workshop, although I could heat it, as I felt the strange surroundings possibly contributed to shock. I put a loose collar and long line on her then set her inside the fenced vegetable garden while I tried to get her fit again. As she lay down, I fed her some warmed lamb milk, left sheep cakes and green garden vegetable leaves by her, then went to fetch a bowl of water. When I got back she saw the blue bowl, struggled to her feet, then made a weak charge almost butting it from my hands. At dusk I set her on a dry bed of hay, then covered her with a blanket. I felt doubtful about her survival. Next morning she too was dead, blood oozing from nose and mouth. Her coat was filled with warble fly larvae.

On my next supply trip an 82-year-old crofter I knew as Sam, who had spent his whole life in the area, came up to me in the store and asked how the deer were faring round Wildernesse.

'You should have many around you now,' he said. Apart from the known fact that red deer come down from high ground in severe weather, I was puzzled how he knew I would have a lot around my small area. 'They like conifer woods mostly, for shelter and browsing,' he said. 'And there's been so much forestry fencing around this region in recent years, thousands of acres, in many places the deer are living in crowded pockets in winter, as they have to come low to find shelter.'

I was surprised to find my own feelings thus confirmed by a man who had known the region all his life. When I told him of the dead and dying deer I was finding, a retired stalker heard our talk and joined in. He confirmed the exceptionally long wet winter was bad for the deer.

'Unless they get six weeks dry after the rut they go right down,' he said. 'Just warble fly in their coats, letting the air in so they get colder and wetter, weakens them, never mind all the other parasites, and the poor feed. And if milk hinds are mistakenly shot it's important to shoot the calves too, or there's a great waste among the calves.' He agreed with me that if the situation continued I should inform the Red Deer Commission.

The situation did continue. On 14 March we found a hind calf had died sheltering under a rock below the second ridge above the cottage. It had been dragged downhill by foxes so it was just a bundle, rump upwards, head twisted in, its two rear thigh bones detached. Two hundred yards further we found a stag calf carcass on a ravine ledge in a fork of the main burn. I felt between its legs – still warm.

A few days later we found a really old hind on the far edge of the west wood. She lay on her stomach, legs splayed out as if she had just made it to the wood, then had collapsed into the final struggle without being able to settle herself first. She had a grey mouth and brown stains between the teeth on her lower jaw. Again blood issued from nose and lips. On 28 March we found another just-dead hind in the east wood, her eyes prominent and bloodshot. It seemed she had been unable to blink towards the end and so had died with her eyes wide open, her eyeballs now completely dry.

The bodies, a few decomposing by this time, were making the woods unpleasant to walk in and there could also have been a health hazard. I perspiringly distributed the carcasses more naturally. I hauled a few up the slopes to about 400 feet for the eagles and buzzards, manhandled two stinking bundles into the raging burn, helping them over snags of rocks with a long pole so the current swept them out to wash up somewhere else. Three I towed away by boat, distributing them evenly along three miles of shore. The remaining few I left where they were.

Next day, as I was at my desk, Moobli came whining to the window. He had found a hind, a young one this time, stuck in a bog on the edge of the west wood. He seemed most concerned at her plight, whining and licking her face as she turned her head away. When I hurried back for gumboots to try to haul her out, he became most anxious, running back and forth, urging me to hurry. She seemed in fair shape, not thin, but the strength in her legs had gone.

I hauled her out by shoving my hands into the mud beneath her chest, wiped her mud-covered muzzle clean, then walked her out of the south-westerly gales to a windless spot behind the cottage where she lay down. Oddly, she seemed little scared. There I set sheep cakes and vegetable leaves and left her alone. Later through the rear window I saw her eating the kale, so I gave her some more. After two hours I checked again – she had gone. I found her sitting fifty yards away in the east wood. I put more food beside her and covered her with a blanket for warmth. By mid afternoon she was standing up, the blanket on the ground. All the food had gone. I gave her more and went back to my desk.

When I returned towards dusk she wasn't there. After a search we found her way over the burn, standing in the wood beyond the bracken flats. As she had chosen, and been able, to walk away from the cottage after feeding I felt her best chance now was on her own, so we left her there. I felt she probably survived.

It was when we found another dead hind calf and then, during a long trek up the big river valley to the east, two dead stags, I decided to write to the Red Deer Commission in Inverness. One stag, a thin old seven-pointer, had been washed downstream and its carcass was tangled up among a jam of branches, old bridge timbers and other debris amid the alder thickets. The other, a young eight-pointer about five years old, did not look starved but had died sheltering away from the prevailing winds behind an ancient ruined wall. I only found it by Moobli's scenting skill. By now the count of dead deer had reached twenty. In my letter to the leading deer expert, Field Officer Louis K. Stewart M.B.E. of the Red Deer Commission, I listed the dead – eight hinds, ten calves and the two stags – and described how the ones I had found dying had distended, slightly bloodshot eyes, made rattling noises in their throats, and that some had blood and a yellowish liquid issuing from their noses, which possibly indicated a respiratory ailment. I said the mortality seemed excessive, though I realised deer did make for woods when sick and that mine were the only sheltering conifer habitat for some miles.

To my astonishment, on the very day I wrote the letter, Moobli and I found *nine* more dead deer – two hinds and seven calves. Again some had blood coming from the nose and most of the calves were very small. When I got back I wrote a second letter to give the R.D.C. the sad new findings. I felt sure some kind of epidemic had broken out amongst the deer. Possibly the greatly increased fencings, involving thousands of acres over the past few years, linked with the low mortality during the five mild preceding winters, had helped the prolific deer stocks build up too high for the ground they had left. Being then driven out of more sheltering glen bottoms they had, along with sheep, possibly overgrazed the hills too. Yet behind the cottage there were still many *unfenced* square miles of range, though here grazing was possibly less nutritious than in the bottoms. It seemed the main factor had been the long cold wet winter.

A few days later we were searching the whole north face of the largest mountain across the loch for eagle eyries when I saw a hind and her calf walking up and down one of the new forestry fences. The calf looked sick, its neck bent back too far. The hind kept poking her head through the wide mesh, and all along the fence the ground had been mulched into browny-black peat by the feet of deer trying to get back to their old grazing grounds. I thought then it was possible the two hinds which had drowned had been trying to swim the loch to reach new ground rather than brave going through the villages at either end.

When Mr Stewart's reply came I felt somewhat reassured that at least there was no epidemic.

We do not often receive such concise accounts of natural mortality and it is a great pity that many more of the deer forest managers do not keep accurate records of natural death amongst deer.

First, the location of carcasses in or near shelter is not surprising. In severe weather deer seek shelter and those in poorer condition tend to stay there. Obviously the shelter described by you becomes the mortuary for the deer stocks in the vicinity.

Natural mortality occurs mainly amongst the oldest and youngest animals and you have already noted the small size of the dead calves. I am sure that most of the adult casualties are pretty old.

The symptoms described by you are probably attributable to severe infections of lung worm and/or nasal botfly grubs. These are common parasites amongst red deer. Lung worms are very debilitating and as the animal gets weaker pneumonia develops, causing death. Nasal botfly grubs are not such a serious problem unless present in very large numbers.

If you care to do some dissection (of the fresher carcasses) both parasites can be located easily. The site of the lung worm is obvious and the botfly grubs can be found by removing the animal's head and splitting it lengthwise. The grubs are about 1"–1½" long and whitish yellow in colour.

Mortality is not uncommon amongst deer at this time of year and there is no cause for alarm unless it occurs amongst prime adult deer . . .

Louis Stewart is the leading authority on Scottish red deer and I felt greatly relieved by his letter. In a phone conversation later he said the twenty-nine deer would have come from a large area and as long as the situation didn't get worse there was no need for an official investigation. I did indeed dissect two calves' skulls and found nasal botfly grubs in both, but after the first dissection of a hind, when I also found lung worm, I realised I was not cut out for animal pathology – I even dislike skinning – so I did no more.

I found three more dead deer over the next few weeks – a hind with a distended gut that appeared 'blown' from eating too much new grass, had fallen down a steep ravine in her weakness, a staggie left draped over a rocky spur by the rushing waters of the burn, and an old hind carcass two miles away. This brought the total to thirty-two dead deer between 22 January and 18 May that winter. Of these, twenty-five were found within a half-mile semi-circle of the cottage (because of the loch); the other seven – three hinds, three calves and the two stags – were found within three miles.

Between 13 March and 7 May the following year, we found a total of eight dead or dying deer: one dying hind and three dead calves in the half-mile semi-circle, and one dead hind and three dead calves on treks within three miles.

The following year, between 19 November and 8 April, we found a total of fourteen: three dead hinds, six calves and a decomposing stag with neck wounds in the half-mile semi-circle; also a dead hind and a calf at 1½ miles, another hind at five miles and a four-year-old stag on the beach three miles away.

Next year, during the hardest winter in sixteen years, between 25 November and 8 April, we found a total of fifteen: seven dead hinds (one with a broken leg after falling from a cliff in March) and four calves in the half-mile radius semi-circle; also two hinds and a calf at three miles, and a sixth calf five miles away.

The year after that, a mild dry winter, between 28 January and 20 February, we only found one dead hind, with another hind dying (she had 143 warble fly larvae in her skin) in the semicircle; also a dead calf a mile away.

It seemed that with the slightly increased culling suggested by the R.D.C., the deer numbers were becoming once again in more acceptable harmony with the environment.

I have saved until the end the most extraordinary experience I ever had with the deer in winter. It brought home to me in a tragic personal manner the winter plight of our red deer and in a way that the mere finding of dead bodies could not. I was no mere observer this time but forced to be an intimate participant.

On 4 April in the year of the main mortality, despite showers of light rain and occasional hail, I decided to boat to the big river valley and then trek up the entire glen to get better photos of stags before they cast their antlers. We headed fifty yards up the fast flowing river, dashed across the current, shut off and tilted up the engine, then rammed the boat up on to tussocks before she was swept back down on to a rocky bank. We trekked up half a mile, crossed the rapidly flowing river on the tops of slippery boulders, and saw two big stags half a mile to our left, grazing just ahead of a spruce plantation.

It would be a difficult stalk; while the wind was from the north-west, the mountains south of the loch sent it curving up from the south-west into the first part of the valley. The land between us and the deer was like a saucer with just a few rocks and scattered young trees for slight cover. We moved on between the alders, turned diagonally across the south-west breeze, then saw an even larger black-looking stag just beyond a rocky ridge a mile to the north-east. There the northwest wind, entering through a gap in the mountains higher up the valley, would be coming true, and I decided to try for this one.

We doubled back half a mile to some drainage trenches then headed east. Crouch-walking through the mud, I looked up once but the stag had gone. Then through the fieldglass I saw it was just sitting down. We had to cover nearly a mile and a half, mostly over rough trenched tussocks to reach a ridge before the one behind which it was resting. I peered to the left: it was still there. As a grey curtain of rain was heading towards us, these next few seconds might be my only chance. I raised up slightly and took two photos of it. At least I had those in the bag, something to show for the day.

There was no shelter when the downpour hit us and we had to wait, soaked and near to freezing in the wind. When the last drops were falling I stalked to the rocky ridge and took pictures of the stag chewing the cud. It stopped chewing, became alert, as if it had heard the last shutter click, got to its feet and trotted a few yards to the side, looking back and giving me a fine frame-filling photo at sixty yards. Then it trotted eastwards and vanished over a crest.

Moobli and I turned west, back to the river, for I wanted to walk up it to the stags' usual main sanctuary in the corrie at its head. We hiked up the east banks as it was easier going and there was no afternoon sun to light us up. After half a mile we crossed a small but noisy burn and went on a further hundred yards. I thought it odd that I could still hear the sound of water splashing. We came round a small spur in the ground to witness a heart-rending scene.

There was a stag stuck in the river!

It tried to stand up, struggling hard as if trapped, then subsided, its nose plunging into the water and out again, holding its head up now as if only with great effort. Once more it tried to get up but stumbled back again into the fast flow. How long had it been there? Maybe all night. Anyway it must have been half frozen. As I walked nearer its head swung round. The pitiful look of terror from those distended eyes as the stag saw our approach was something I shall never forget. Helpless, pathetic, he lay there like a stag at the end of a long chase with deerhounds, when the men with guns arrive to end its life.

He was thin, his ridged spine sticking out distinctly from his back, and his stomach bulged as if blown from eating the new grass. Yet this was a big stag with a wide sweep of antlers and weighing well over 200 lbs. Part of me wanted to leave him so that he wouldn't struggle through fear, but I had to see what was wrong and try to save him if I could. I couldn't ford the foaming river at that point. Keeping Moobli to heel, I managed to cross a double fork higher up, then came back down the far bank.

As I came near, the stag again tried to struggle to his feet but the rear right foot seemed to be wedged between two boulders under the water. This would not have held a stag normally, so he was clearly ill. As I side-stepped down the bank he attempted to gouge me with his antlers trying to stick and sweep inwards with the tips. I managed to grab the tip of one antler and in his unnatural position, lying over boulders, he could not exert much force. He got up on to his forelegs and again tried to hook but I managed to hold the tip away from me and then jump back. Still he could not pull the leg free and subsided weakly. I waded into the icy foaming flow behind him, grabbed his right fetlock below the water after pushing up my sleeve and heaved his leg down and back. On the third pull, as I tried to say calming things to the stag, the leg came free. He had very big feet I noticed, and was bald on his crown, with grey ears and not much hair on his muzzle. His antlers were wide but 'going back', for they were poorly tined with only seven points. He was an old beast for sure. Even now he couldn't get up and he just lay there, his head resting on the bank, bloodshot eyes popping, as if resigned to his fate at human hands, a fate that was no doubt bred right into his instincts.

The longer he stayed there the colder he would become. I *had* to get him out somehow. I clambered on to the bank, emptied my boots, squeezed my socks dry, then put them all back on.

'Come on, old lad,' I said. 'This is it. We're going to help you, get you out of there.'

Getting as firm a footing as possible, I took hold of both antlers with what I hoped were encouraging words, and bit by bit heaved gently – no sudden pulls for I was afraid of breaking his neck. The stag got his two front feet up and tried to walk with them. He seemed to be trying to co-operate but the right rear leg had either gone to sleep or was near frozen, for he seemed unable to use it. Luckily it was a gentle slope and I finally hauled him right out, but he just lay there, eyes rolling, too weak to move further. He closed his eyes, half opened them again.

It was terrible to see him there, his long graceful legs stretched out, his flank heaving, the great spread of antlers distorting the lying position of his head and pushing his muzzle into the ground. The fallen monarch. How many glens, how many miles had passed beneath those flying cloven hooves, how many peaks had he gained to see a world few humans ever see, how many harems had he herded together, hinds had he served, defended against other stags? How many times had he roared his challenge across the glens in the fall? How had he felt in his youthful prime when proud of his strength, proud of his speed, he had gazed with fearful disdain at the stalkers labouring below, for by heaven he had evaded them all? But now Time, the greatest stalker of all, had caught up with him, as it does with us all, the foolish, the clever, the weak or the strong. I could only guess what was passing through his mind as he lay there, humbled in old age. I felt a surging sadness as I looked at him. How little we really do for these beautiful creatures with whom we share the web of time and life. I felt a strange identification with him too, both of us trying to survive alone in the wilds – he in far harsher conditions than I faced all the time, I to try to express a vision that might yet lie beyond my powers, both fighting a losing battle with time and the elements.

I wouldn't leave him like that, lying helplessly on his side. I went round to his rear and pushed him upright. As I reached under his stomach to pull his lower leg up so he would be lying naturally, he swung his head round and his right antler just missed my eye. I clutched the antler at the tip where he would have least leverage and forced his head against his body, then with all four feet correctly placed, I jumped back as I let go. Then I left him. He gazed at our departure in weak surprise for he must have thought he was going to be killed.

We trekked to the end of the valley and saw several more stags feeding. When we came back he had gone. I hoped he would survive but I doubted it. As we tramped wearily down the river to the boat I felt less sad. While professionally I could think he should have been taken out by a bullet a year before to save the waste of his probable close death, that was a concept of waste only from the human view. At least this way he would have lived to real old age and a natural death. He had had his day – and a full one at that, and none of us can hope for more.

As I walked back along the woodland path, the still wintry light glinting from the loch upon the trunks of the alders, Moobli wending his powerful gentle bulk as quietly as a cat behind me, I thought how much more we could do for the deer, and for all our wildlife generally.

As winter slowly turned into spring, and spring into summer, I was reminded again, as I am year after year in the wild places, of nature's eternal cycle of renewal if enough habitat is left alone to evolve naturally. In June I visited the site of every winter carcass which had first given vital food through the harsh months to eagles, buzzards, ravens, crows, foxes, wildcats and even shrews. Now there seemed more to the old adage 'from dust to dust, ashes to ashes', for around each one the herbage was lusher, greener than elsewhere. My tallest sweet chestnut tree, grown to ten feet in five years, had been planted where a hind had died the previous winter. In nature there is no waste. All flesh is grass and back to grass goes the flesh. The deer had returned to the earth, the bodies had bestowed their minerals and nutrients to the soil, and now were dissolving away so almost nothing but for a skull or backbone could be seen any more. And between them the flowers glorified the sun. Now the summer leaves were grown I found scores of new rowans, birches and a few oaks had seeded in the west wood. Because the deer numbers had been so reduced, some of the tips of the oak trees in the front pasture had also escaped their browsing. Indeed, four of these oaks had overtaken the ones I had bought from outside the area and had planted myself.

At the time I still felt sure the deer herds had suffered overall from the many deaths, on top of the normal cull of stags and hinds. But on 19 September, hearing the first roaring of a stag, we trekked high up the eastern end of Guardian Mountain, came over the last ridge cautiously and there before us was a magnificent sight. A vast herd of 206 red deer, all hinds with their yearlings and calves, were filing slowly through a fold in the hills, looking like a great herd of caribou in the Arctic tundra in the days of yore. And there, as yet walking peacefully among them, their antlered heads held high, were the first two master stags of the year.[*]

There was as yet no danger. The weak had fallen but high in the hills, licked by the flame of the grasses turning gold and the yellowing bracken, the balance that ultimately triumphs in the way of the wild, the great cycle of creation itself, was being renewed.

[*] The Deer Commission for Scotland, while pointing out the difficulty of accurately counting deer in forest environments, estimated populations of red deer increased from 197,600 (plus or minus 35,000) in 1967 to 350,900 (plus or minus 33,300) in 2000. The most recent population estimates for Scotland for all four species of deer in 2013 suggest overall numbers of between 360,000 to 400,000 red deer, 200,000 to 350,000 roe deer, 25,000 sika deer and 2,000 fallow deer. They point out it 'remains clear that the range of all four species of deer in Scotland has been increasing'.

PART SIX

Chapter 21

Renewal

I have lived in this wild place for ten years now; it is my eighteenth year in the wilderness with only boat access to my remote homes. And while the last winters have seemed a little harsher, the boat heavier and the hills steeper, each season still brings the reassurance and joy of new discovery.

On a sunny day in early June in our seventh year Moobli and I were out on the Hill looking for new born red deer calves to photograph. Up the steep tussocky sides of the burn's ravine we toiled, out above the woods bordering the loch, then over undulating slopes of peat bogs and heathery plateaus to just above the big corrie at 1,600 feet.

Suddenly a large broad-headed auburn dog fox, with dark grey patches moulting on his coat, shot up a few yards ahead and lolloped slowly up the steep hill. As I slid the camera pack from my shoulder, Moobli surprised me: for once he chased after the fox without command. I knew he had no wish to hurt it, but just wanted to catch it up and corner it for me if possible. Besides, Moobli had developed an arthritic growth in his rear left leg and, no longer fast, he ran with a limp.

As usual when a fox is chased, this one loped tantalisingly ahead, just one mile an hour faster than the dog. As I fitted the long lens, the fox reached the crest, looked back at Moobli with what seemed a wide grin of disdain, then vanished over the top as my shutter clicked. I caught up, told Moobli off in no uncertain terms, then set him to track the fox to its den. The scent ran east over the tops, then south and back down into the woods below.

We were supposed to be after calves and I had no wish to reclimb the Hill, so I called him off.

We found our first dappled youngster, a tiny stag calf, lying in a grassy hollow. Usually they are quite trusting but, as I lifted this one's short tail to confirm its sex, it let out a loud screech, ran totteringly for a hundred yards and flopped down again in a small peat hag.

Then I recalled the fox. It had not been lying out in the daytime open just to sun itself, but had probably been spying for calves. This calf had now let loose far more scent by moving than if it had stayed where it was. My fault. I walked with Moobli in a large semi-circle, knowing the fox would be reluctant to cross our scent, certainly not before the calf 's mother returned from grazing to feed it. Luckily, I could see her, watching us anxiously from the near skyline.

We tracked round the far side of the mountain, a full two-mile circle, photographing three more calves without touching them, before I realised we were coming back to where the fox had jumped up. I knew the fox to be an intently curious creature. If it thinks it hasn't been seen, it will lie doggo amid the herbage and watch you go by! For this reason, and also because it is likely to be near its den, it will also circle back to where it had been before being disturbed. Had this dog fox done that?

We moved carefully upwind, scrutinising every brown or reddish patch of dried moss or old bracken. Then as I peered round a sharp ridge of rock I saw a small orange patch in the grasses. Moobli was scenting strongly but also obeying his training to keep back. Painfully slowly, I edged the camera round, *Click*. It *was* a fox, a lovely long-legged vixen. She jumped on to a boulder to give me a superb sideways picture, staring, with intent suspicion in her dark orange eyes. Next she dropped her forepaws on to a lower rock, swung her rear legs round as daintily as a cat, and vanished amid the boulders.

We crept forward, and then heard a loud thumping in the den below. The vixen was telling her cubs to scatter and squeeze themselves into rock crevices. A mature fox, she well knew that the approach of human and dog could mean terriers sent down to kill the cubs – just one of the methods used by Highland farmers and keepers to control foxes.

I whistled unconcernedly as we left. After half a mile we doubled back to a high point to keep watch. Two hours later, at dusk, there had been no movement. Two days later we returned – the den was empty. As usual, even after slight human disturbance, this vixen had moved her cubs elsewhere.

Earlier in the same season, some strange droppings had begun to appear in our little woods. Hard, often containing the blueblack wingcases of beetles, these scats had twisted tapering ends and were left on mossy rocks, old stumps and even on the top rock step between my log gateposts. For weeks I assumed they were being left by either a very small fox or a young otter, or even a large stoat.

One mid-June day we were walking through the east wood by the shore when Moobli gave an odd huffing snort. Instantly a dark brown animal leaped from a stump and rapidly climbed up a thick larch between the full foliaged beech trees. It looked like a huge chocolate-coloured squirrel but a gale was blowing and because the trees, twigs and leaves were constantly shifting I couldn't see it clearly. Only when it reached the top, then stared down at us and I saw its creamy throat patch, did I realise what it was – a rare pine marten! And it was the first I had seen anywhere near Wildernesse.

For months I tried experimenting with different baits to get it to where I could take a good photo, but without any luck. Then, in the autumn of the following year, I got my chance. Just before dusk I had left out on the bird table two slices of bread and margarine, which had fallen briefly into the wet sink and become partly soaked. Suddenly, as I sat at my desk, I was aware of a large dark animal sitting on the window sill, its bushy tail brushing the pane. It was the marten again. He jumped on to the table, grabbed one slice of bread in his jaws and leaped to the ground with it.

From then on I set out bits of bread and jam every night, and a few nights later I got my first pictures by flashlight. He was a fine-looking beast, a full 30 inches long, with lustrous fur, hairy padded long-clawed feet and big dark eyes. He became tamer and tamer. We had many adventures with that pine marten over the next few years, but they belong in another book. As do the stories of three remarkable owls, a wounded badger that needed help, and four young foxes.

On another early June day I lay in a superbly camouflaged hide overlooking my thirtieth eyrie across a ravine, almost overcome by the finest golden eagle experiences I had yet known. The young mother was exceptionally devoted to her chick, and during my first 21-hour visit she spent all but half an hour on the nest with it – feeding, preening, tucking it in or just gazing lovingly at it!

This pair had brought in, among their normal prey, two well-grown fox cubs, a large roe deer fawn, an old heron, a merganser, a stoat, one adult and two young ravens and several hooded crows – all classed as vermin if you're a shepherd, forester, fisherman or poultry-keeping crofter.

Today the smaller sleeker male landed beside the female as if begging her to leave the chick and come away for a long float together. She glared at him with piercing orange eyes. He appeared to receive a message. Off he flew, to return with a bare stick which she took from him, dropped as if in disgust, then gave him another meaningful look. Off he went again, this time returning with a great spray of rowan leaves. This she weaved into the nest to screen her bairn from sun and wind. Twice more the male returned with leafy sprays, until she had almost completely covered the eaglet from my view. Only then did she fly away with him.

After leaving the hide in the early hours next morning, I had only gone 300 yards down the hill when I heard the chick calling out. Instinct made me drop my pack, take out my camera and click on the telephoto lens. It was lucky I did, for just then the mother eagle came winging round a cliff – and I got one of my best flight photos of these glorious Highland birds.

My eagle observations of earlier years are the sole subject of a book I published in 1982 under the title *Golden Eagle Years*. I have worked a great deal more with eagles since and hope to bring that story up to date before too long.

I am too tired to slog up the mountain to the eagle eyrie again today, so I do a little gardening, and then go to sit with Moobli for a restful half hour above the loch and between the little woods. The summer blue sky, the boundless arch of heaven, hangs above us like a great bellflower, bestowing beauty and light over the earth, while above the mountain slopes shimmer in the hazy heat. I look at the front pasture, now revealing its full splendour among the thick fescues, grasses and wild flowers that have flourished from my yearly labours to cut back the engulfing bracken. Humble bees drone amid the tall purple spires of foxgloves. Bright sparks of butterflies flicker everywhere – blues, small coppers and heaths, speckled woods with lemon spotted wings, gaudy orange fritillaries, and rare little checkered skippers whose numbers have quadrupled now their food plants have increased. With their swift darting flight they are hard to see above the golden buttercups, the blue speedwells and the white trellis works of the pignut flowers. One lands on an early ragwort, holding its wings stiffly open in the sunlight, as if to reward me with a good photo.

Suddenly a flashing movement among the carpet of bluebells which seem to reflect the blue vault of the sky below the crimson flowers of a rhododendron bush. It is a rare broad-bordered bee hawk-moth, long thought to be extinct in Scotland. As it pendulums from flower to flower, probing a long proboscis into each frilly bell for nectar, its thick bee-like body seeming to hang without support because its mainly transparent wings are invisible in flight, I feel a great thrill. Here is even more proof of the success of cutting the bracken and encouraging honeysuckle to which many handsome hawk moths go for nectar, and on which this scarce moth's caterpillar feeds. A small bright bird flashes in from the west wood – a male siskin, rarely seen at Wildernesse, with brilliant yellow patches in his green plumage. With its drabber mate it happily feeds from the reddish seeds of dock plants I have also encouraged, because of their tasty spinach leaves.

Moments later there is a flash of bright pink winging over the flowers and I am seeing my first magnificent painted lady butterfly here. She flies over to the purple rhododendron by the path, meets another, paler in colour, and both twist up in spirals through the hot air, glittering pinks, whites, buffs and greys, about to mate. As I hope the female will lay eggs on the small bed of nettles I have planted specially for her kind, or on the thistles I have left to bloom, I am more astonished by a splash of deep yellow among the flowers. It is a clouded yellow butterfly which is not supposed to occur as far north as Scotland. I hope it too will breed in the red and white clovers flourishing near my garden fence.

From the middle of the loch I hear mewing wails of a rare black-throated diver, swimming along, perhaps to relieve its mate before their two eggs finally hatch on the islet where the gulls decorate the lichened rocks, as white as golf balls in the distance. Overhead the female buzzard beats westward, the sun reflecting from the loch waters burnishing the colours of her underwings.

For a year I have seen no roe deer but this morning we saw a buck and his doe in the west wood. We withdrew quietly leaving them to graze in peace before they headed up the higher slopes in the sunlight. Up there, beyond the ridges where the eagles, buzzards and kestrels hunt, the red deer hinds which still winter in our woods have had their first dappled calves, now sitting patiently in shady bracken or heather until their mothers return from grazing to give them milk.

Before us now a large *aeshna* dragonfly hawks below the trees, blue body and gauzy wings glittering like a fairy helicopter, the dashing Red Baron of the summer skies. I hear the waters of the burn, reduced to a trickle after a long hot spell. After a gap of a year we found this morning two new otter spreints on the mossy rocks lining its banks, showing that a new young animal has moved into the area, just as a new young heron has taken over poor Harry's ground.

Beside us the woodland shade is filled with flitting forms, bird song, and the wheezy squeaks of youngsters where tits, chaffinches, thrushes and robins are feeding bairns. Willow warblers and stonechats have brought their young to feed on the tribes of insects above the increased wild plants of the pasture, and a pair of spotted flycatchers have adopted our south-eastern garden post as their favourite perch, making aerobatic dashes, hoverings and rumblings in their pursuit of flies. I hear a scuttering sound and look up to see one of the red squirrels darting along the reddish scaly bark of a Scots pine.

All seems at peace at Wildernesse; forgotten yet again are the dark cold days of winter for now the whole world is aflame with light and warmth. Quite suddenly I feel almost overcome by the beauty around me, the sweet scents, the gently swaying trees, the murmurs of the breeze, the soft lap of waves against the shore, the orchestra of bird song, the humming of bees and tiny insects wings, the tinkling of the burn etching its way over clean washed stones. Over all there is a pervading harmony, a glimpse perhaps of a world of balanced beauty in which what we, in our varying ways, call God meant man and all wild creatures to dwell, a glimpse indeed of paradise on earth.

In nature's teeming world the animals and birds are working hard to fulfil their destinies. The feeling came strongly upon me that we, who evolved from original creation to become the dominant species, with unique gifts of intelligence, foresight and the ability to love spiritually beyond ourselves, have an inherent and inescapable duty to act as responsible custodians of the whole inspiring natural world. We are the late-comers, it can only be ours on trust.

If we let it down then we also let down its Creator; and even if we don't believe in God, conservation of the natural world and its ability to inspire our finer thoughts – for only thought can change the world – is without any doubt whatever a necessary ethic of our own survival. The kingdoms of the wild evolved in creation not for mere man to plunder, to satisfy greed under the guise of progress, and finally to destroy, but both to enjoy and enhance. If we fail to learn from the last wild places, we may yet create a hell on earth before we too pass along the road to extinction, the fate of all dominant species before us. Spiritual unease has long been manifest. The lessons will not wait forever to be learned.

Also available by Mike Tomkies

Between Earth and Paradise
An Island Life

After giving up a hectic life as a journalist in Europe and Hollywood in the late 1960s to return to his boyhood love of nature, Mike Tomkies moved to Eilean Shona, a remote island off the west coast of Scotland.

There he rebuilt an abandoned croft house and began a new way of life observing nature. He tracked foxes and stags, made friends with seals and taught an injured sparrow-hawk to hunt for itself. It was the indomitable spirit of this tiny bird that taught Tomkies what it takes for any of us to be truly free. Whether he was fishing, growing his own food or battling through stormy seas in a tiny boat, he learned that he could survive in the harsh environment.

This is the astonishing story of daring to take the first step away from urban routines and embracing a harsh yet immensely rewarding way of life which, in turn, led Tomkies to an even more remote location and inspired an acclaimed series of books on various animals and the challenges and joys of living in remote places.

ISBN 978 1 78027 706 6